Patrick Moore's Practical Astr

For other titles published in the series, go to
www.springer.com/series/3192

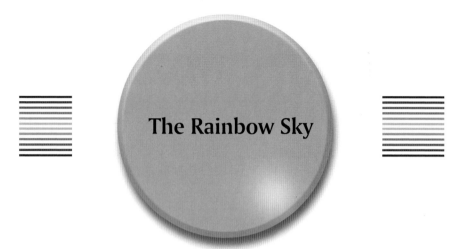

The Rainbow Sky

An Exploration of Colors in the Solar System and Beyond

Tony Buick

Tony Buick
Orpington, Kent
UK
tonytz@waitrose.com

ISBN 978-1-4419-1052-3   e-ISBN 978-1-4419-1053-0
DOI 10.1007/978-1-4419-1053-0
Springer New York Dordrecht Heidelberg London

Library of Congress Control Number: 2009939497

Printed on acid-free paper

Springer is part of Springer Science+Business Media (www.springer.com)

*To my dear, beautiful Caitlin, who became
1-year old during preparation of this book.*

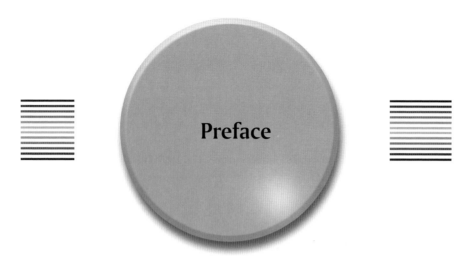

Preface

Where do you start to write about colors in the universe? Do you look to the deepest ocean trenches on Earth, with their awesome bioluminescent creatures roaming the blackness of the abyss? And where do you finish? With the most distant galaxies in the cosmos? A difficult question, perhaps, but in between the two extremes, there is so much to marvel at that it really doesn't matter where you start or end, as long as you note the staggeringly beautiful and complex examples of color there are and that each should, if possible, be represented in some way. Whether staring up at the sky when surprised by the sudden appearance of a vividly colored band of light that is a rainbow or peering through a telescope to view colors further afield, the origin and complexity of the source of light is witness to the wonderful and majestic world and the universe in which we live.

An attempt has been made here not only to create a picture gallery of the universe, but also to provide brief explanations or interpretation of the colors and, where appropriate, to give hints on how to capture pictures easily yourself, without spending lots of money. As illustrated in the introduction, paying attention to just a few basic camera settings, it is possible to turn a blurred snapshot into a detailed and pin sharp picture worthy of framing and hanging on the wall.

Much effort and research has been put into ensuring that the facts, figures, and anecdotes in this book are correct. There are, of course, the inevitable conflicting stories, facts, and data. Some technical descriptions or historical facts differ where they shouldn't, as they describe well established knowledge. Some are clearly silly slips, such as one astronomical poster that quoted the diameter of the Sun that is actually the radius. Some differences are quite understandable, as theories and hypotheses develop and change. In an attempt to resolve such differences, where possible, several sources of information for each fact were sought and studied to determine the correct or most likely explanation or true value, which entailed plowing through a great many books and Internet websites.

One objective of the research was to ensure that the contents were up to date, but therein lies a problem. New discoveries occur almost every day. New and stunning images are beamed back to Earth from the planetary and interplanetary missions, in addition to those from the ground-based and space-based telescopes, with such overwhelming frequency that a completely up to date treatise is impossible. However, perhaps that is of little consequence, if the aim is to showcase natural and false colors that represent objects and events throughout the Universe, maybe even

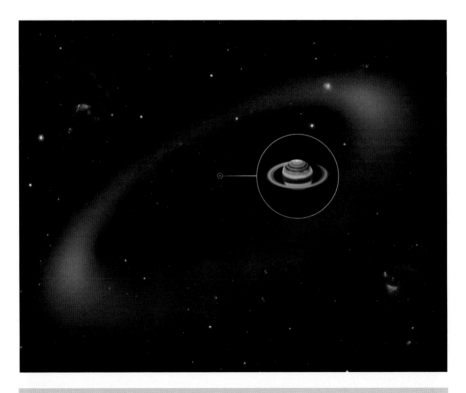

Fig. 1. An artist's representation of the ring (Credit: NASA/JPL-Caltech/Keck.)

capturing a few of the images yourself along the way. As if to emphasize the constant discovery of unimaginable space objects, a huge ring (red representing an infra red image) was discovered by NASA's Spitzer Space Telescope, in October 2009, around Saturn (see Fig. 1).

I do hope you enjoy reading this exploration of color as much as I have enjoyed preparing it.

Orpington, Kent, UK   Tony Buick

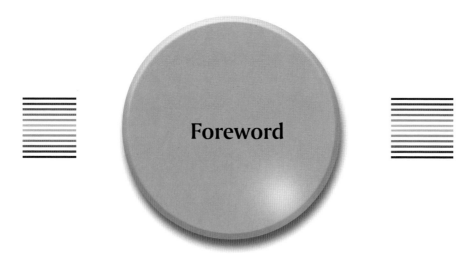

Foreword

We live in a colored universe, and perhaps we do not always appreciate its variety and beauty. In his new book, *The Rainbow Sky,* Tony Buick has collected images of many objects from many different sources, each bringing out particular aspects and hues of events or phenomena that can be seen above our heads. For example, there are Earth artifacts, rainbows, Sun dogs, aurorae, planetary scenes, vividly colored nebulae and galaxies, and much more.

Some of the images have been taken by amateurs; others were taken by giant Earth-based or space-based telescopes. There are pictures that were transmitted by orbiting vehicles, providing a window on spectacular galaxies thousands of millions of light-years away.

It is clear that great care has been taken over the selection, and the effort has been well worthwhile. All the pictures are well captioned, and there is also an accompanying text that will be appreciated and savored even by readers who are not artistically inclined.

Tony Buick has captured many of the photographs himself and has given practical instructions and advice on how anyone can do the same, often with simple equipment. So the book is practical as well as informative.

There is much to learn about the universe, and this book provides a fascinating tour with which to introduce the subject as well as to explain why the stunning colors in publicly available space pictures are not always what they seem.

This is a very unusual and comprehensive book, a most enjoyable work.

Sir Patrick Moore

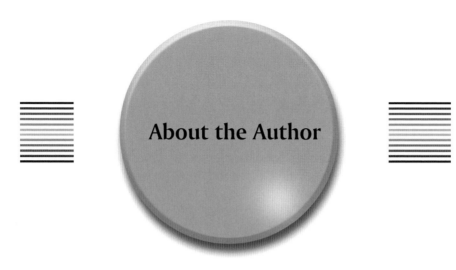

About the Author

Although Tony Buick in his career worked in medical, veterinary, and agricultural science, specializing in analytical chemistry, he turned to his lifelong interest in astronomy following an early retirement and has encouraged the younger generation to observe and understand the sky while teaching science, computing, and geography. His fascination with the Moon was given a further boost through his friendship with Sir Patrick Moore, which led to the publication of *How to Photograph the Moon and Planets with your Digital Camera,* now going into its second edition. Tony Buick has a wide range of interests, from the "infinitesimal" under a microscope to the 'infinite' through a telescope and has published articles on tardigrades, the robust microscopic animals that can even survive for a while in space, in addition to articles on the Moon. This book is a product of Tony Buick's interest in spectroscopy and color in general throughout the universe.

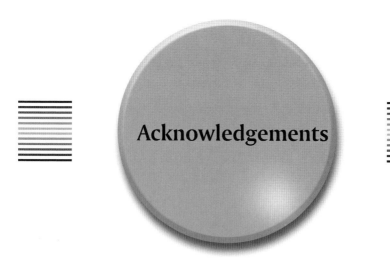

Acknowledgements

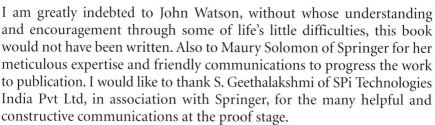

I am greatly indebted to John Watson, without whose understanding and encouragement through some of life's little difficulties, this book would not have been written. Also to Maury Solomon of Springer for her meticulous expertise and friendly communications to progress the work to publication. I would like to thank S. Geethalakshmi of SPi Technologies India Pvt Ltd, in association with Springer, for the many helpful and constructive communications at the proof stage.

Gilbert Satterthwaite and Jeff Harries undertook the major task of reading through the manuscript, for which I am extremely grateful.

Where pictures are provided by others, credit is given, and it is a testament to their generosity that so many have allowed reproduction of their images within this work, either after direct contact or by making their images freely available on the Internet, especially NASA and ESA and the picture website flickr.com. It has also been most enjoyable communicating with so many friendly and talented people around the world during discussions of use and origin of images. Interpreting clouds and colors in the atmosphere can be tricky, and I am grateful to the UK Meteorological Office for their expert replies to my many questions.

Members of the Orpington Astronomical Society have been most encouraging and permitted some astonishing images to be reproduced here.

The support by family and friends has been invaluable, and thanks again to my sons and their families – Tim and Chris (images and computer support), and Kat and Jo. Thanks to Eileen Thompson for belief in the work and practical assistance and Wendy Poole and Roy Bareham for constantly listening with enthusiasm to my reports of "how the book was going."

Finally, I owe so much to Sir Patrick Moore for his encouragement and friendship over many years that made such a difference in my astronomical pursuits.

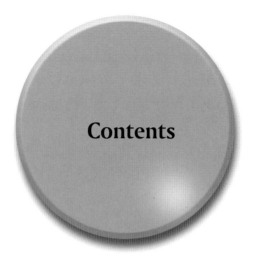

Contents

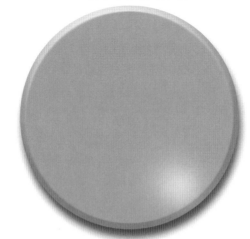

CHAPTER ONE

Introduction

We take for granted the vivid colors that are all around us, and that is how it should be. Who wants gray roses? Wonderful shades of green abound, punctuated by splashes of reds, blues, and yellows, and stunning combinations of these colors provided free by nature in town and country. Artificially created colors can be pretty, too, such as those from inert gas lights – if we can forget the boring messages they often blurt out! Imagine if there were only shades of gray. No! Let's not! Most will now be thinking of trees and flowers, city lights, and art galleries. That is a two-dimensional perception, forward and back, side to side, perhaps a few meters of depth, with much to appreciate and investigate (Fig. 1.1).

The third dimension for colors sometimes thrusts itself into view suddenly and unexpectedly. Quite common is the awe-inspiring blanket of a bright red sunset covering the entire sky, often developing into a multitude of other shades of red, yellow, blue, and even green (Fig. 1.2).

And why? The same reason as why the daytime sky is blue – Rayleigh scattering owing to air and dust molecules being much smaller than the wavelengths of sunlight. All light is scattered, but blue light is scattered more than red light. So, during the day, predominantly blue light is scattered downwards to make the sky appear blue, but in the evening, when light has a much longer pathway through the atmosphere, blue is

T. Buick, *The Rainbow Sky*, Patrick Moore's Practical Astronomy Series,
DOI 10.1007/978-1-4419-1053-0_1, © Springer Science + Business Media, LLC 2010

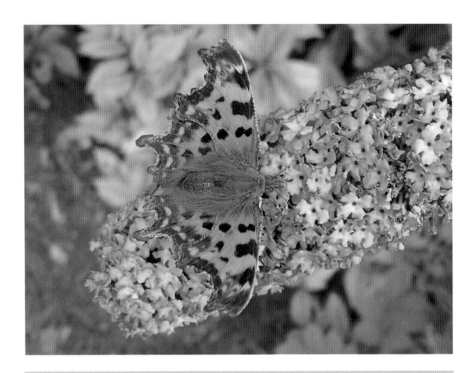

Fig. 1.1. Nature's colorful world, a Comma butterfly on a Buddleia bush.

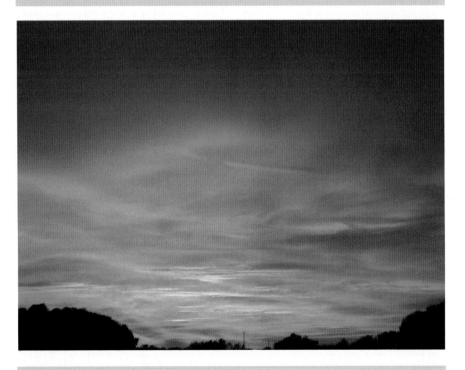

Fig. 1.2. The scattering of blue light leaves an awesome red sunset.

scattered away (down to the distant daytime part of the world, leaving their sky blue), allowing red to dominate the evening sky. Atmospheric haze and volcanic particles (and the perception of color by the human eye) all have a part to play, but the most intense sunsets are produced by clean air and appropriately placed clouds. See the Appendix in this book for more information about this.

Rainbows are ephemeral and beautiful, dazzling one minute, gone the next. However, they do illustrate the range of colors visible to the human eye – red, orange, yellow, green, blue, indigo, and violet (see Fig. 1.3).

Sunlight traveling through each of the multitude of raindrops is refracted, reflected, and then refracted again. With multiple internal reflections higher orders of bows appear.

Just as common, if you are looking for them, are Sundogs, or Parhelia, that grace 22½ degrees either side of the Sun (see Fig 1.4). Maybe you will even see a full circle, a halo, if you are lucky. These lovely patches of Sun dog color are produced by light shining through high altitude ice crystal clouds. With a little camera know-how you can create wonderful photographs of them.

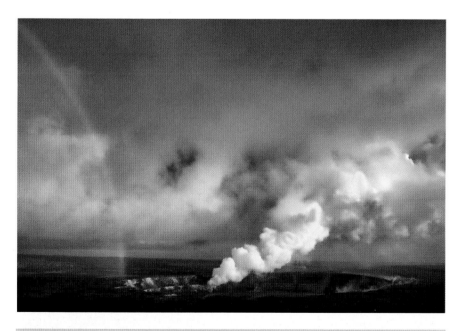

Fig. 1.3. A rainbow over the eruption of Halema at Kilauea. (Credit: Mila Zinkova, http://home.comcast.net/~milazinkova/ Fogshadow.html.)

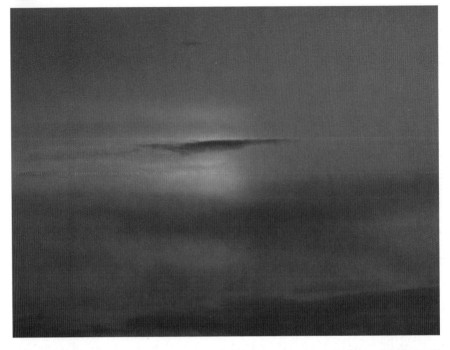

Fig. 1.4. A Sun dog in a warm summer sky.

Photographing sunbeams or crepuscular rays (see later) are within the ability of all, although practice is required to attain super images, such as in Fig. 1.5.

Rarely, a whole sequence of events gels into one experience, as the orientation and shapes of the colors change over a few hours. For more on this, see Chap. 8.

Common in certain parts of the world are breathtaking aurora, mostly sighted in the Arctic (Aurora Borealis) and Antarctic (Aurora Australis) regions (see Fig. 1.6). David Cartier, the person who photographed this picture, writes: "Believe me; the vocal harmonies of a wolf pack and the dance of the aurora make a sweet combination."

Much less commonly seen, let alone photographed, is the phenomenon called zodiacal light. Zodiacal dust consists of tiny particles, remnants of comets and asteroids, that reside in the ecliptic plane and reflect sunlight and are seen soon after sunset and before sunrise. The reflected light is very faint, and the skies must be very dark to capture the glow. An impressive image is shown in Fig. 1.7 with an exposure of

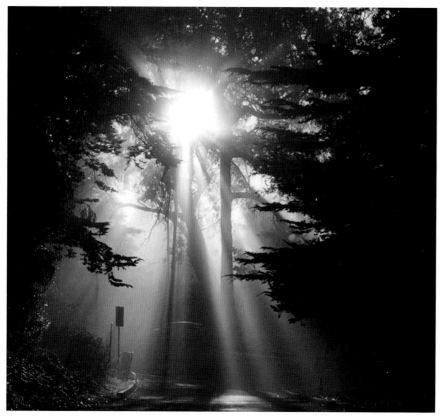

Fig. 1.5. Brilliant rays stream through the trees. (Credit: Mila Zinkova, http://home.comcast.net/~milazinkova/Fogshadow.html.)

a few minutes that also illustrates how the colors of stars are brought out by spreading the point source of illumination, known as star trails. More on star trails later.

We do not often even consider looking high up in the sky when searching for colors, for many reasons. A person must first be aware that they are even there; then an effort has to be made to access them. If capturing your own pictures from space is not your ambition, then astronomy books and the Internet are superb sources for such pictures. But to experience the thrill of capturing your own unique color image you need to beg, borrow, or even purchase equipment. As the distance to the colorful object increases so might the time and/or cost of the equipment required for the production of reasonable resolution images. Within reach of the amateur are images of the Sun, Moon, planets, galaxies, nebulae, and stars

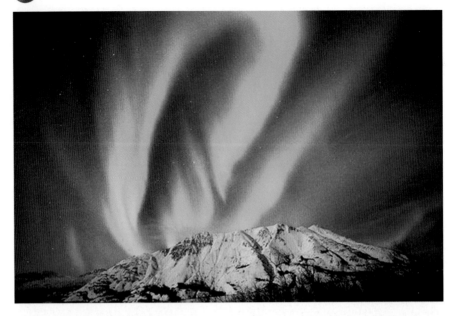

Fig. 1.6. Aurora, Sheep Mountain, Yukon. (Credit: David Cartier.)

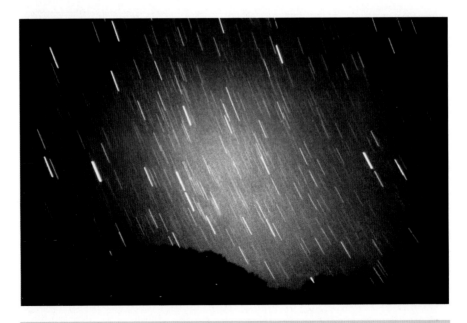

Fig. 1.7. Zodiacal light in Gemini photographed under the dark skies of Tenerife. These are simple examples of great colors found just above – relatively speaking – ground level. (Reproduced with kind permission from Dr. Brian May.)

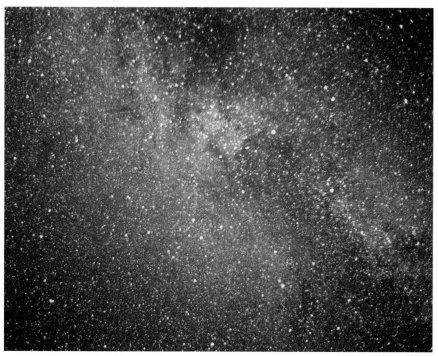

Fig. 1.8. The star field in Cygnus. (Credit: Paul Whitmarsh, Orpington Astronomical Society.)

(see Fig. 1.8). Each of these will be covered in the following chapters, where identification, celebration, and explanation of astronomical colors will feature more prominently than methods of capture. However, there are some great and very simple ways to gather astronomical colors for fun, and these will be described in more detail later.

False colors – colors assigned to cosmic objects and phenomena to graphically represent their properties – can often mistakenly be understood to be real among the plethora of space images constantly being published (see Fig. 1.9). Important science is often hidden within a true-color picture that is sometimes produced merely for visual enjoyment. Not until we actually visit the source of those pictures, Mars for example, will it be known as to how close those images, generated to be true color, will be to what is actually seen. Figure 1.10 provides a great example of true and false color. The colors in the picture were enhanced to bring out the real difference in color between the stone and the surrounding granular spherules. Even true color photographs often begin their life as a collection of color-filtered monochrome images. The label "representative" is sometimes preferred, accompanied by explanations for the assignment

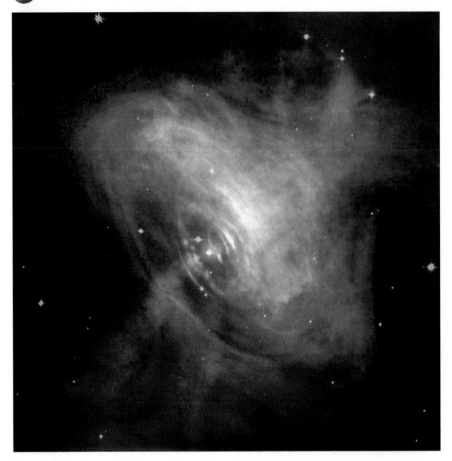

Fig. 1.9. Combined X-ray and optical images of the Crab Nebula. (Credit: NASA/CXC/ASU/J and HST, Hester et al.)

of particular colors to radiation received originating from distant physical and chemical processes. And since the human retina varies from person to person it is not possible to create a true color image that would appear the same to all. More on this, later.

Whatever the reason for the colors they are wonderful pictorial representations that help us understand and be amazed at the fascinating universe in which we live.

But also note that the best details are often forthcoming from images in gray scale, or black and white (see Fig. 1.11).

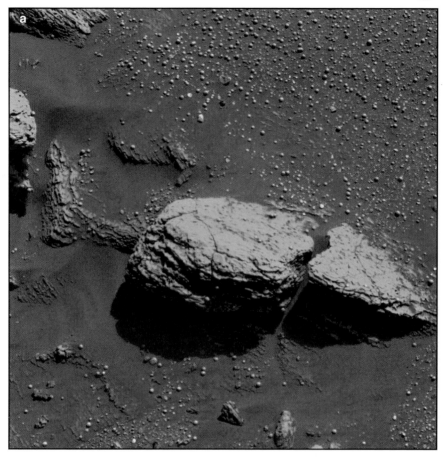

Fig. 1.10. The same Mars rock imaged in true and false color by the rover Opportunity. (Credit: NASA/JPL/Cornell.)

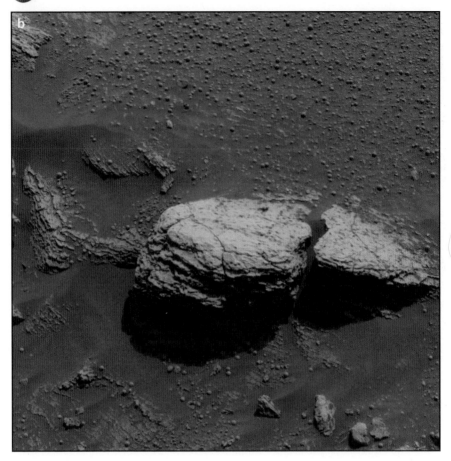

Fig. 1.10. (continued)

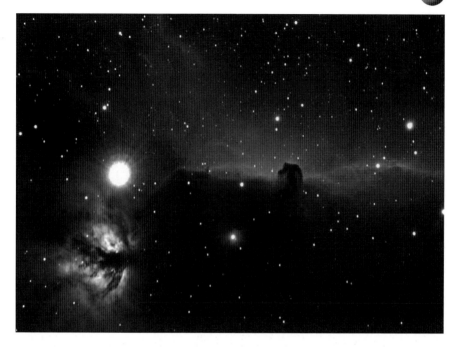

Fig. 1.11. Horsehead Nebula, IC434, *black* and *white* image using hydrogen alpha filter only. (Credit: Mike McRoberts, Orpington Astronomical Society.)

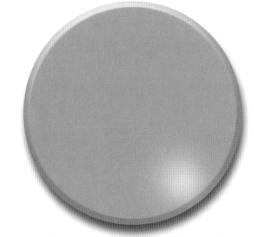

CHAPTER TWO

The Origin of Light

T. Buick, *The Rainbow Sky,* Patrick Moore's Practical Astronomy Series,
DOI 10.1007/978-1-4419-1053-0_2, © Springer Science + Business Media, LLC 2010

Fire

Where does light come from? Not that many people would think to even ask this question, let alone know the answer. If asked, many people would guess that the answer is "fire."

Early humans must at least have been aware of fire (see Fig. 2.1), with an occasional loincloth being singed by an unexpected streak of lightning! And there is evidence that at least as far back as 400000 BC. Peking Man was aware of fire. Clearly, burning stick torches were used by the Cro-Magnon (28000 BC) and to light the dark caves in order to produce paintings such as those at Lascaux in southwestern France, thought to be 16,000 years old (see Fig. 2.2).

Shell lamps have been found with grease (animal fat) and wicks (13000 BC), and shells and indentations carved out of walls served as reflectors. Saucer lamps have been unearthed dating from around 1000 BC (see Fig. 2.3). Oily birds and fish threaded with a wick (5000 BC) were made into lamps (Yuk!). And the many different oils and fats made the lamps

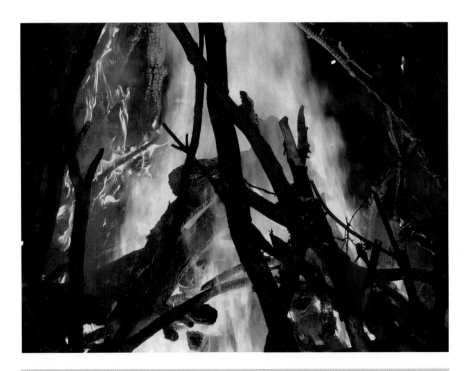

Fig. 2.1. Fire, the original source of man-made light.

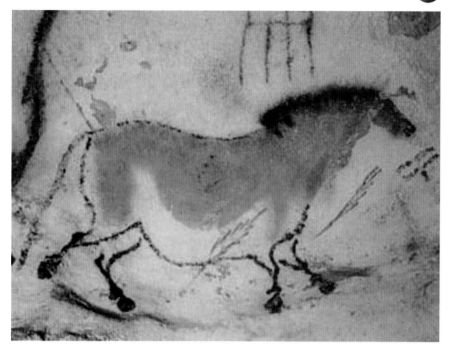

Fig. 2.2. Image of a horse from the Lascaux caves.

edible in times of famine or hardship. Shakespeare, in his play, *The Comedy of Errors*, suggested threading an overweight kitchen maid with a wick to produce a candle that would burn for a very long time!

An early approach to thinking about the actual structure and composition of light was that of the very famous mathematician, Pythagoras (582–500 BC), who theorized that every visible object gives out, in straight lines, a stream of particles that hits the eye, and that the light was traveling from the eye to the object. Plato (427–347 BC) stated that the effect of seeing occurred when the light arrived at the object. Heraclitus (535–475 BC) hypothesized more broadly when he suggested that the world was created neither by god nor man but was, and is, fire being ignited and extinguished regularly and eternally. Aristotle (384–322 BC), a pupil of Plato's, promoted a new theory that light travels like waves. He also considered music to be something akin to colors where the most enjoyment came from the arrangements of notes into chords and the arrangement of colors, respectively.

Lychnology is the study of ancient lamps and a lychnoscope is a small side window in a church that was used for lepers (and others) to be able to see the altar.

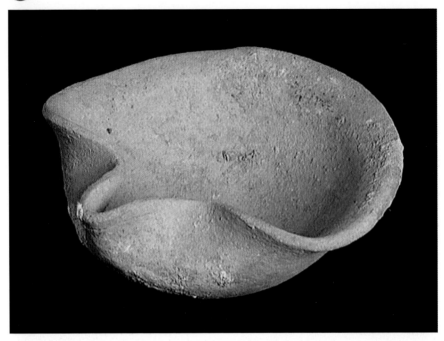

Fig. 2.3. A pinched terracotta saucer lamp ca. 1200–800 BC. (Credit: Edgar L. Owen, Ltd.)

Light and optics were studied by scholars like Euclid (320–275 BC), who postulated that the speed of light must be very high. "You close your eyes and the stars are gone, you open your eyes and they immediately appear, even the distant ones!" An interesting bit of logic! Fire was still the source of light in the Roman lighthouses (pharos) from 280 BC (see Fig. 2.4), for Roman theaters (55 BC to AD 200; see Fig. 2.5), for horn lanterns (AD 100), and for candles (from AD 400). Possibly the greatest pharos of all was the Great Lighthouse of Alexandria in Egypt that was around 500 ft high (approximately 150 m) and one of the Seven Wonders of the Ancient World. Built in 270 BC on the island of Pharos, the name of the island became strongly associated with lighthouses.

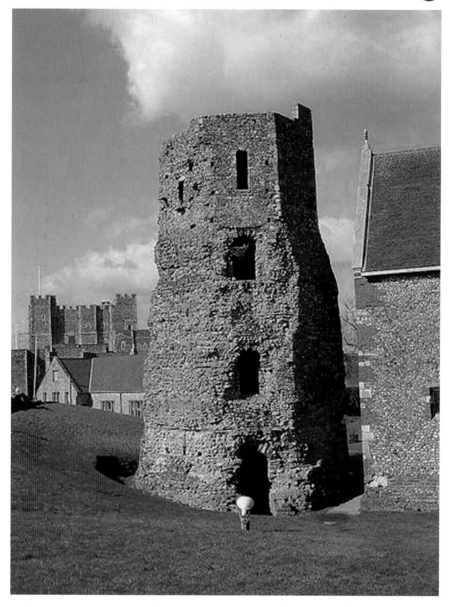

Fig. 2.4. First century AD Roman Pharos at St Mary-in-Castro, Dover Castle, England. (Credit: Wayne Hopkins, TX, USA.)

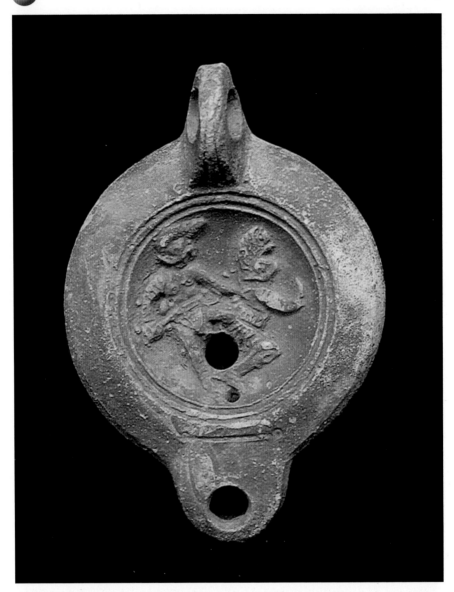

Fig. 2.5. Roman gladiator lamp, ca. First and second century AD. (Image courtesy of Edgar L. Owen, Ltd.)

Gas and Electricity

Sebastiano Serlio (1475–1554) investigated color filters and, using a bottle and a brightly polished barber's basin behind a candle, produced a spotlight. Shakespeare reported that candle wicks needed to be trimmed constantly. Other sources of light were found to be interesting. Robert Boyle (1627–1691) observed light from bacteria and fungi (bioluminescence) and Hennig Brand discovered light-emitting phosphorus obtained from human urine (1669). The magic lantern (1650), ancestor of the modern slide projector, also appeared about this time (see Fig. 2.6).

Of course, it must not be forgotten that fire, in the form of supplied town gas, was still being used for lighting well into the twentieth century. At the start of the Victorian era most houses were illuminated by candles and oil lamps. Although gas lighting was introduced in the streets of London as early as 1816, it had not come into general use until the Houses of Parliament started using it in 1859, which demonstrated to all that it was trustworthy. There was a lack of trust presumably because a number of gaslit dwellings had been accidentally reduced to ashes! Large pendant fittings boasting several burners were known as "gasoliers," some of which were extravagantly ornate (see Fig. 2.7). Fish-tail or circular (Argand) burners spread the flame to optimize light output. Then, in 1897, the incandescent gas mantle was introduced, which was adapted to burn downwards and send the light to where it was needed instead of always at the ceiling.

Not until electricity was discovered and recognized for what it was (early humans were surely familiar with the bolts out of the blue, and the ancient Greeks knew electricity through rubbing fur on amber) and harnessed by famous scientists like Benjamin Franklin (who was lucky not to have been electrocuted by his kite aerial) and Michael Faraday (creating the electric dynamo) were fires gradually replaced by artificial sources of illumination. Thomas Edison's first viable incandescent light bulb lasted 45 h, beginning in the afternoon of October 21, 1879.

Coal, via coal-burning power stations, has been a major source of electricity since the nineteenth century, and fossil fuel burners (thermal power stations) still outnumber nuclear, geothermal, biomass, and solar thermal plants around the world. The heat of the burning fuel is converted to steam energy, then mechanical energy to drive generators to produce the electricity. Coal or town gas was originally a by-product of the process of coking, and other by-products such as ammonia and tar were useful to the chemical, feedstock, and dye industries. Typical composition of town

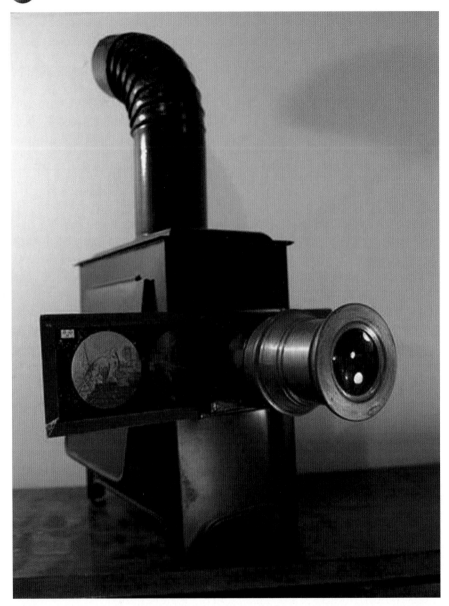

Fig. 2.6. This superb magic lantern once belonged to the photographer's great grandfather. (Credit: Andrei Niemimäki.)

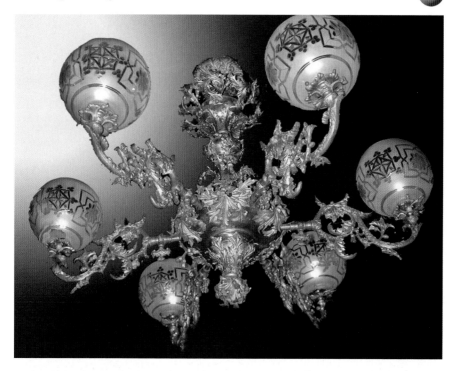

Fig. 2.7. Valuable and beautiful six-arm gasolier measuring approximately a meter wide and a meter high. (Image courtesy of C. Neri Antiques and Lighting.)

gas was hydrogen, 50%, methane, 35%, carbon monoxide, 10%, and ethylene, 5%. The switchover to natural gas in mid-twentieth century was precipitated by the demonstration that it could be transported safely, economically, and efficiently over long distances. It was also cleaner but required complex processing to remove impurities and produce the useful by-products of sulfur, ethane, and natural gas liquids such as propane and butane. Around 95% of piped natural gas is methane.

However, the familiar and ubiquitous daylight does not come from fire – or does it?

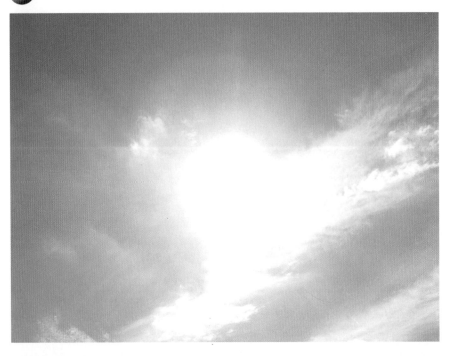

Fig. 2.8. Just visible in this image is a colored ring around our great ball of fire in the sky. Note that the colorful ring that appears in the image is not due to an atmospheric phenomenon but to a common distortion of the Sun's image through a camera lens.

To state the obvious, it is dark at night and light during the day, when the Sun reappears. So our light comes from the Sun, which looks like a great ball of fire (see Fig. 2.8).

The UK meteorological office wrote the following about the Sun in October 2008:

> One of the main problems when photographing the Sun via the lens of a camera is the amount of diffraction and distortions caused by the sunrays going through the glass. This is a case here with the Mickey Mouse ears that appear in the photograph. Having said that, studying the image in closer detail there does indeed seem to be some kind of optical phenomena associated with the Sun in this image. It's certainly not a halo as the rinds are far too close to the sun for this to happen. My guess is that what you have here is a corona. The description that appears in the Meteorological Glossary best describes this phenomena, and I've quoted this below:

Corona – a series of colored rings surrounding the Sun or Moon. The space next to the luminary is bluish-white, while this region is bounded on the outside by a brownish-red ring, the two forming the 'aureole'. In most cases the aureole alone appears, but a complete corona has a set of colored rings surrounding the aureole – violet inside, followed by blue, green, and on the outside yellow to red. The series may be repeated more than once, but the colors are usually merely greenish and pinkish tints.

The corona is produced by diffraction of the light by water drops, Pure colors indicate uniformity of drop size. The radius of the corona is inversely proportional to drop size; thus growth of a corona indicates decrease of drop size.

A corona is distinguished from a halo by its reversal of color sequence, the red of the halo being inside, that of the corona outside; the dull red, which is the first notable color in the aureole, ranks as outside the bluish tint near the luminary.

Nuclear Energy

Fire is normally recognized as being the combustion of materials in air or oxygen to produce heat and light, and the Sun has predominantly hydrogen with a lesser amount of helium and much smaller quantities of other elements. A nuclear reaction, hydrogen being converted to helium, causes the emission of heat and light. Much research has been carried out by scientists to determine how the Sun works. Many a theory has fallen by the wayside, but as more and more probes are launched into space to complement ground-based instruments in observing and studying solar and other cosmic emanations, the mechanism for the continuing existence and functioning of our very own star is becoming clear.

Nuclear reactions as a source of power used by humankind, for both useful and destructive purposes, are now very familiar (see Fig. 2.9). For the production of huge amounts of energy there are two types of reactions, fusion and fission. Power stations for the generation of domestic electricity, propelling nuclear submarines, and all other nuclear reactions harnessed by humankind utilize fission (see Fig. 2.10).

Fission is a decay process in which an unstable nucleus splits into two fragments of comparable mass. Following earlier work by Enrico Fermi, Otto Hahn and Fritz Strassman hurled neutrons at uranium and discovered previously unknown isotopes of barium and krypton among the fission fragments. Over 100 other nuclides representing 20 different elements were also in the resultant mêlée. Massive amounts of energy in the form of heat and light were also produced, far beyond the energy released from any known chemical reaction. In science-speak two isotopes of uranium, the common (approximately 99%) ^{238}U and less common (approximately 1%) ^{235}U (others are also known) are found in nature that undergo fission through bombardment by neutrons. Fast neutrons are required to activate uranium 238, and slower ones work for uranium 235. The reaction also *produces* neutrons, so a chain reaction can occur to fuel power stations if controlled or an explosion if uncontrolled. Only a brief description is given here because fission is not the nuclear reaction that powers the Sun. The one we are interested in is nuclear fusion.

Whereas fission breaks up large atoms into smaller ones, fusion creates larger atoms from small ones – in fact, the smallest. Since the time of the Big Bang, hydrogen atoms, H, and molecules, H_2 (two atoms bonded together) have dominated and still dominate the composition of the universe. We say "from the time of the Big Bang," but at the very beginning, around 13.7 billion years ago, no atoms or molecules could possibly exist in the outrageously unimaginable temperatures of many billions of

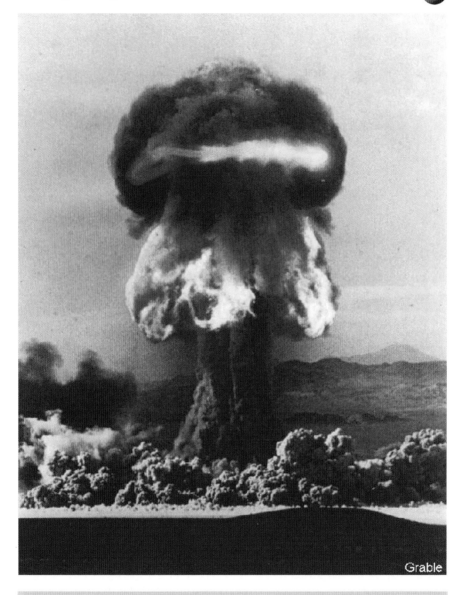

Fig. 2.9. Nuclear bomb test in Grable, Nevada, in 1953. (Credit: Gregory Walker, Trinity Atomic website.)

degrees Kelvin (or Celsius, the number of degrees of which are exactly the same, but zero starts at around 273° higher). Only quarks, subatomic particles, were flying around, unable to stick to anything at such huge energies. Only after a brief time (but very long time in early cosmological terms),

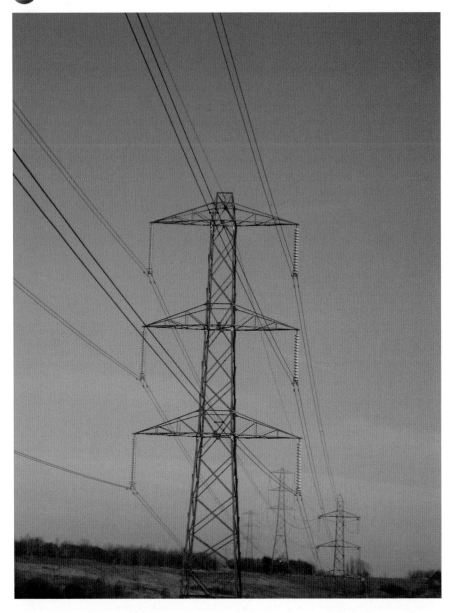

Fig. 2.10. Here electricity produced from nuclear fission flows through cables for peaceful purposes.

a microsecond, was it possible for quarks to settle down into forming nucleons, protons, and neutrons, and then into forming hydrogen and helium, although still rushing about at enormous speeds and without associated electrons, which were too fast to capture. Most figures associated with description of the goings on at that time are staggering. Astrophysicists struggle to theorize what happened prior to 10^{-43} s, known as the first Planck time. Quantum theory is king when considering interactions at the atomic level, where energy is not continuous but comes in discrete bundles (quanta) and time in definite intervals. Hence the first Planck time. Energies on the grander scale are described by Newtonian physics, and maybe, one day, theoreticians might just combine the two. Much effort is being put into discovering such a unified theory acceptable to all.

For the next 300,000 years there was a gradual cooling until, when the temperature reached about 30,000°, electrons were slow enough to be captured by nucleons to form neutral atoms – the negative charge on the electron balancing that of the positive proton. Disturbances in the system allowed atoms to clump together to form massive stars that quickly burned out, exploded, and spread the debris to form aggregates elsewhere. Smaller and longer living stars evolved. It must be borne in mind that modifications to theories of processes occurring around the time of the Big Bang are constantly being suggested, such as the proposal that at the end of inflation (a sudden expansion postulated to account for the present structure of the universe), 10^{-32} s, the maximum temperature was "only" 11,000 K!

The subject of categorization of stars is taken up in a later chapter. But here we will look at why and how stars employ nuclear fusion to stay alive.

The baryonic universe, that is, the matter that we can see and measure, accounts for a very small portion of what is believed to be the total content of matter, perhaps only a few percent. Hydrogen contributes to the largest quantity of such matter, often found in huge clouds that pervade the whole of space. Cosmic events, perhaps pressure waves from explosions, trigger the coming together of bunches of hydrogen atoms that then, through gravity, attract other atoms. Increasing amounts of matter, and therefore increased gravity, lead to a rise in pressure at the center. High pressures generate high temperatures and, when high enough, a nuclear fusion reaction is triggered and a star is born. Massive Jupiter, with its gaseous content, is often referred to as a failed star, but it would have to be very much bigger (more massive) to reach the minimum mass to generate the required pressures, and hence temperatures, to initiate a stellar nuclear reaction.

An atom of any element consists mainly of space. Only a tiny nucleus resides in the center, and an even smaller electron flies around it or, in quantum mechanical terms, has a probability of being somewhere around it. The nucleus is composed of two basic entities – protons, with a single positive charge, and neutral neutrons, which are strongly bound together. The proton is 1,836.12 times the mass of the electron. Hydrogen is unique in having only one proton and no neutrons, at least in its simplest form, although neutrons are incorporated into its other forms, or isotopes. All other elements possess neutrons and protons.

At the temperatures deep inside the heart of a star, around 15 million Kelvin, protons cannot hang on to their electrons, which move around with no loyalty to any particular nucleus. The hydrogen is therefore an ion, a positively charged nucleus, without the negatively charged electron that would effect a balance of charges. Two other facts need to be stated to enable a simple set of equations to be written to describe fusion. Firstly, a proton can lose its positive charge through ejection of a positron, a positively charged electron, to form a neutral neutron. Secondly, a particle called a neutrino that has no charge and is almost weightless can carry away energy to make the equations balance.

Neutrinos are so small and insignificant that they fly through atoms hardly ever interacting with one. Huge numbers even travel through your body having traveled a direct route from the other side of Earth!

In the mêlée of the stellar furnace the following particles are involved in fusion: protons, deuterons, neutrons, neutrinos, and positrons.
The simplified reactions are

1. Proton + proton → deuteron + positron + neutrino

    $$^1H + {}^1H \rightarrow D + e^+ + \nu_e$$

2. Deuteron + proton → helium (light helium-3) + gamma ray (energy)

    $$D + {}^1H \rightarrow {}^3He + \gamma$$

3. Helium-3 + helium-3 → helium-4 (normal helium) + proton + proton

    $$^3He + {}^3He \rightarrow {}^4He + {}^1H + {}^1H$$

There is a tiny difference in mass between the starting protons and the final helium nuclei and protons. However, the small loss in mass has been converted into enormous amounts of energy, according to the famous Einstein equation $E = mc^2$, which is responsible for the heat generated to fuel the star.

The gamma rays are photons that eventually reach the surface of the star to be emitted as radiation, some of which appears as the visible light we see.

The Sun "burns" around 400 million tons of hydrogen every second.

So nuclear fusion is simple, isn't it? Hydrogen ions stick to other hydrogen ions to form the larger helium nucleus and some energy in the form of heat and light. Not so simple, though, because there are energy barriers to overcome for the particles to get close enough to do more than dance around each other. For protons to fly into each other to achieve the proximity required for a reaction the speed would have to be huge. At the center of the Sun, where the temperature is 15.6 million Kelvin, the speed is only $6.2 \times 10^5 \, \text{ms}^{-1}$, much lower than that required to overcome the electrical repulsion between two protons. A proton would have to be at a temperature of 11 billion Kelvin for the nuclear fusion to happen.

However, nature has some wonderful fudge factors for getting around the seemingly impossible. The Tunnel Effect, to overcome the energy barrier, is one such, where particles can also be considered as waves, with the tiny probability of existing outside their local environment to overlap, and hence combine with, other particles. The above equations to illustrate the mechanism of heat, light, and chemical generation within the Sun necessarily present a summary of what is going on. Further explanations require a deeper delving into the wonderfully weird world of quantum mechanics.

It should be quite apparent to all that the world around us is not just hydrogen and helium. An essential player in the stellar inferno to enable promotion of helium to higher atomic masses is beryllium-8, an unstable isotope with an extremely short half-life of $2.6 \times 10^{-16} \, \text{s}$ and decaying back to helium-4. As more and more helium is produced it combines with hydrogen and more helium to produce unstable entities that then revert to helium. As the core of the star begins to collapse the center's temperature rises to about 100 million Kelvin, at which point there is just enough concentration of beryllium-8 and helium-4 to convert to stable carbon-12 and oxygen-16, the building blocks of life. For stars with the appropriate temperatures other reactions are possible to produce even higher atomic masses.

Within the stellar furnaces nuclear reactions continually force entities together and blast them apart to create not only heat and light but also heavier atoms and, eventually, all 92 naturally occurring elements. The periodic table shown in Fig. 2.11 systematically displays all the natural elements plus a few additional ones made by us, some of which are very short lived. To achieve the heavier natural materials it is necessary for several generations of stars to explode and re-form, involving additional

Group →	1	2	3	4	5	6	7	8	9	10	11	12	13	14	15	16	17	18
↓ Period																		
1	1 H																	2 He
2	3 Li	4 Be											5 B	6 C	7 N	8 O	9 F	10 Ne
3	11 Na	12 Mg											13 Al	14 Si	15 P	16 S	17 Cl	18 Ar
4	19 K	20 Ca	21 Sc	22 Ti	23 V	24 Cr	25 Mn	26 Fe	27 Co	28 Ni	29 Cu	30 Zn	31 Ca	32 Ge	33 As	34 Se	35 Br	36 Kr
5	37 Rb	38 Sr	39 Y	40 Zr	41 Nb	42 Mb	43 Tc	44 Ru	45 Rh	46 Pd	47 Ag	48 Cd	49 In	50 Sn	51 Sb	52 Te	53 I	54 Xe
6	55 Cs	56 Ba		72 Hf	73 Ta	74 W	75 Re	76 Os	77 Ir	78 Pt	79 Au	80 Hg	81 Tl	82 Pb	83 Bi	84 Po	85 At	86 Rn
7	87 Fr	88 Na		104 Nf	105 Db	106 Sg	107 Dh	108 Ils	109 Mt	110 Ds	111 Rg	112 UUb	113 Uut	114 Uuq	115 Uup	116 Uuh	117 Uus	118 H

Lanthanides	57 La	58 Ce	59 Pr	60 Nd	61 Pm	62 Sm	63 Eu	64 Gd	65 Tb	66 Dy	67 Ho	68 Er	69 Tm	70 Yb	71 Lu
Actinides	89 Ac	90 Th	91 Pa	92 U	93 Np	94 Pu	95 Am	96 Cm	97 Bk	98 Cf	99 Es	100 Fm	101 Md	102 No	103 Lr

Fig. 2.11. Periodic table of the elements. (Credit: Cepheus, Wikipedia Commons.)

reactions. Stars larger and hotter than our Sun can use carbon, if already present, as a catalyst for an alternative route for the "burning" of hydrogen. A star really is the alchemist's ultimate dream of a factory to change elements into other elements. See the Appendix in this book to find references that can be used to pursue this topic further.

Looking up into the sky on a dark night encourages us to ponder on all those nuclear reactions producing light, without which the universe would be a very gloomy place indeed. So stars and stellar reactions are the source of the light that we were seeking – fires of a sort, but not quite the wicks and oils, gas, and incandescent features of places here on Earth.

CHAPTER THREE

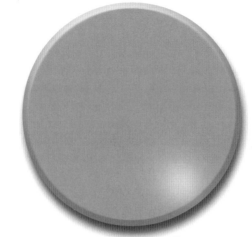

Properties of Stars

T. Buick, *The Rainbow Sky,* Patrick Moore's Practical Astronomy Series,
DOI 10.1007/978-1-4419-1053-0_3, © Springer Science + Business Media, LLC 2010

Star Mass

Stars are hot bundles of gases – our Sun being the nearest, of course (see Fig. 3.1) – that emit huge amounts of light and exist because of their furnace of nuclear reactions. But the reactions are not all exactly the same, and the lifetimes and characteristics of the bundles are not the same, either. To complicate matters even further stars mature and change their characteristics, and when they explode they encourage next generation stars to whip up more complex products from the original ingredients. So it is necessary to consider the various types to account for the properties of light from particular stars that reveal the existence of a multitude of elements in our observable universe, not just hydrogen and helium.

The amount of light or radiation emanating from stars (their luminosity) varies enormously from thousands of times fainter than the Sun to millions of times more. However, most star masses are from 0.3 to

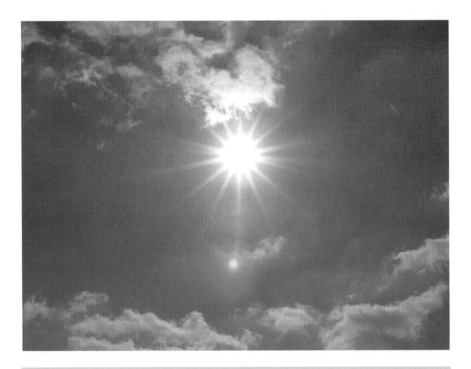

Fig. 3.1. Our Sun on a bright summer's day.

3.0 times the mass of our Sun. There is currently some uncertainty as to how big a star can be. Theoretically, masses can be as great as around 120–150 times the mass of the Sun. The upper limit is based on the fact that temperatures generated are higher with greater masses until a value is reached where the temperature cannot be sustained without the entity just exploding. Plaskett's star, in Monoceros, was discovered by J. S. Plaskett in 1922 and was found to be the most massive binary star known. Each of the two stars weighed in at about 50 solar masses. There are a few stars, such as Eta Carinae, that may have masses of more than 100 times that of the Sun. Bodies smaller than 0.08 times the Sun just would not have enough mass/pressure to initiate a nuclear fusion reaction to get things going. A mission is planned, SIM PlanetQuest, that, if approved, will provide the breakthrough technology needed to determine these two extremes of stellar evolution.

> Luminosity, or absolute brightness, is the amount of radiation emitted by a star. The Sun's luminosity is 3.846×10^{26} W.

Not to be confused with luminosity is magnitude, which is the brightness of a star. Ancient Greek astronomers defined the brightest stars, those first visible as darkness descended, as those of first magnitude. Hipparchus, around 120 BC, cataloged 850 stars, rating their brightness on a scale of 1–6, 1 being the brightest. With the development of instruments for the measurement of the brightness of stars, it was recognized by Norman Pogson in 1856 that each magnitude was different from the next magnitude by a factor of approximately 2.5. He subsequently realized that five magnitudes were roughly equivalent to 100, then redefined magnitude to be exactly this. Each magnitude difference was then 2.512, or the fifth root of 100. The zero standard star he chose was Polaris, but when it was realized that the Pole Star was slightly variable, Vega was chosen to take its place. Modern astronomy now calculates magnitude using different reference factors and Vega is now 0.03. More on this later.

> The magnitude of a star is its brightness. The apparent magnitude is as we see it from Earth. The absolute magnitude is its brightness as it would be from a standard distance away, 10 parsec (1 parsec = 3.2616 light years).

The mass is also responsible for how long a star lasts. In fact, the mass is responsible for just about everything to do with a star. Large masses give rise to higher temperatures at the core that accelerate nuclear fusion and therefore consume the fuel (mass) at a faster rate. Big stars die young and small ones can potentially have lifetimes in excess of the age of the universe. Higher temperatures lead to a greater output of light that is referred to as its luminosity. Hotter stars are more luminous. When mention is made of light output what is actually referred to is radiation, some of which cannot be seen by the human eye. Before continuing toward the modern classification of stars the composition of "light" and its mechanism of generation must be addressed.

For a thorough understanding of light and atomic structure, references are given in the Appendix of this book, but not that much information is required for a basis on which to describe the variety of stars that we see.

The Electromagnetic Spectrum

Once Newton separated sunlight, normal white light, through a prism into colors it opened a whole new way of interpreting what is seen (see Fig. 3.2). It became clear that light is just one type of energy that is radiated throughout the universe. Visible light is a small part of what is known as the electromagnetic spectrum, so called because the waves have an electric and a magnetic component. The stunning beauty of the colors is easily produced using a CD (compact disc), to bounce direct sunlight onto a white background (see Fig. 3.3). The closely spaced lines on the CD act together to separate (diffract) the light (see later for the explanation of diffraction).

The full spectrum consists of electromagnetic radiation from very short wavelengths to very long (see Fig. 3.4). Many of the names of sections of the spectrum will be familiar, such as X-rays, ultraviolet rays, and microwaves, indicating the wide variation in the properties of

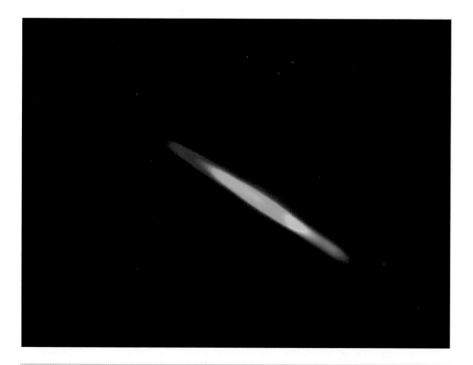

Fig. 3.2. The wonderful colors of white light are revealed through diffraction.

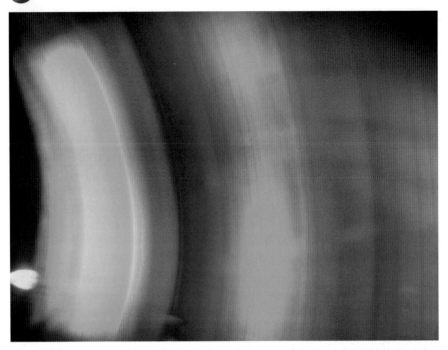

Fig. 3.3. Beautiful colors from sunlight are produced through diffraction from a CD.

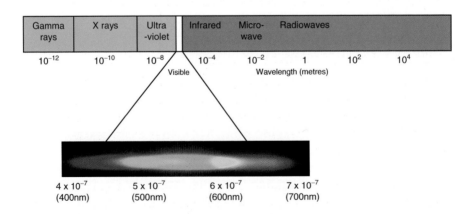

Fig. 3.4. Here is the full electromagnetic spectrum.

different wavelengths. Although the range is large, the human eye can only see a small part; wavelengths of between 0.4 and 0.7 micrometer (μm), 400–700 nanometers (nm), or from blue to red.

Wavelengths represent energies. In the diagram the lowest energy is the radio wave and the highest is gamma radiation. All types of EM (electromagnetic) radiation can be described as consisting of photons that have no mass and travel at the speed of light. The position on the EM spectrum informs as to how much energy those photons have.

Radiation from every part of the EM spectrum is constantly flying around all through space, but, with the unaided eye, only the white or colored light can be seen. All objects in the universe give out, reflect, or transmit some light or radiation. The composition or profile of such emanations is entirely characteristic of that object. Profiles can be of a narrow band of wavelengths or a combination of wavelengths that tell the exact state and makeup of the object when the radiation left that object. It must be remembered that since light travels at around 300,000 km or 186,000 miles per second some heavenly bodies, millions of light years away, at the time of observation will have aged enormously since the body emitted that light or may no longer exist. Even the Sun is 8 min older by the time its light is seen here on Earth. Colors represent energy. Heat a material that does not burn, and the hotter it gets the more red it gets, then white. The white heat has more energy.

Joseph von Fraunhofer was born in Straubing, Germany, in 1787. His brilliance at making the world's finest glass was the envy of even the likes of Michael Faraday. He invented diffraction grating and incredibly precise methods for measuring dispersion. He died young at the age of 39, in 1826, possibly due to heavy metal poisoning, as did many glassmakers of his time.

Although in the present context colors are of prime interest, many spectra have extra information. They can contain not just continuous colors but also dark lines at particular wavelength positions – first discovered by Joseph von Fraunhofer in 1814 – that are characteristic of the elements causing them. Dark lines are seen in the spectrum of a star when the light passes through a cooler layer of outer gas that absorbs radiation emitted from the hotter gas below. Bright lines on their own are produced in emission spectra due to the release of a photon from an atom when a previously excited electron jumps from a higher to lower energy level within that atom. Fluorescent lights are typical emission spectra sources.

Atomic Spectra

Much has been written about spectra and what can professionally be achieved. However, it is possible to get your own spectral photographs for fun and to illustrate the EM radiation by using very simple equipment and techniques. First, though, it might help to run through the fundamentals of why spectra and dark and bright lines occur, starting with an introduction to atomic structure for complete beginners that seasoned astronomers will wish to skip. Again, references to more in-depth studies are given in the Appendix of this book.

The smallest part of an element, any element, that can exist and maintain its chemical properties, is an atom. Its size is unbelievably small. A speck of material the size of a pinhead will contain trillions of atoms. One atom of helium is around $0.000,000,000,1$ (10^{-10}) meters in diameter.

> The number of atoms in the observable universe is estimated to be about 10^{80}.

Although it is very small it has been known for some time that the atom consists of even smaller particles. Results of early experiments led to visualization of an atom as a central mass of two particles, protons and neutrons, clumped together, with electrons whizzing around in an orbit at some distance, relatively, from the central clump, called a nucleus.

The proton carries a positive electrical charge and the electron a negative charge. Atoms can consist of many protons and electrons, but in their normal state of existence the number of protons must equal the number of electrons to maintain an electrically neutral entity. Figure 3.5 represents a neutral hydrogen atom.

Electron: 9.11×10^{-31} kg
Proton: $1,836 \times$ electron, 1.67×10^{-27} kg
Neutron: very similar mass to proton
Proton and neutron diameter: 2.5×10^{-15} m

The type of material, or chemical characteristics, of an atom are determined by the number of protons. For example, hydrogen is hydrogen because it has one proton. Carbon has 6 protons, iron has 26, and each material, or element, has the same number of electrons to achieve electrical neutrality. Let us ignore neutrons for now.

For the moment, let us go with this model of the atom, as it helps to illustrate the physical processes associated with an element's behavior. We will see later that it gets much more complicated.

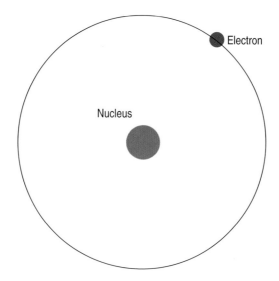

Fig. 3.5. A representation of a hydrogen atom.

An electron in an orbit is clearly in an energy state that keeps it just the right distance for survival without flying away from, or being attracted by, the proton in the nucleus. If more electrons are added to balance more protons in a different element they cannot all crowd into the same orbit. They have to find orbits of their own. This is a fundamental law of atomic structure that also decrees how many electrons can be in each orbit and how much energy each orbit represents. Orbits may hold two electrons each. Carbon, therefore, may be represented by our simple model of dots and circles, Fig. 3.6.

Each of the circles or orbits has its own energy; the further out the higher the energy. If left alone, the electrons and nucleus will happily remain in this configuration. However, if a particle of radiation, a photon, hits one of the electrons with the right amount of energy, then the electron can be knocked into an orbit of higher energy, and some radiation, or light energy, will have been absorbed. This absorption can be detected. The higher energy state of that electron, its excited state, might be unstable and drop back to its normal or ground state (see Fig. 3.7). Some light will then be emitted. The absorption and emission energies are characteristic of the element, hence the reference to absorption and emission Fraunhofer lines.

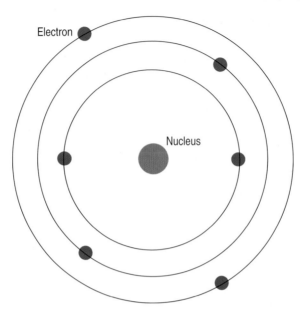

Fig. 3.6. Simplified representation of a carbon atom.

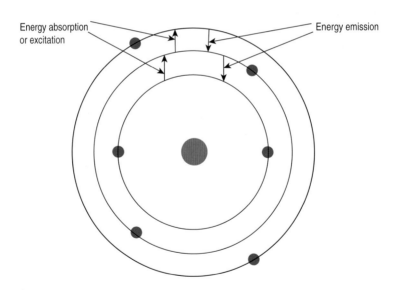

Fig. 3.7. Electron transfer between atomic energy levels.

It may appear that there might be an infinite number of energy states, depending on how far away the orbits are from each other. That is not the case, however, and each distance is a definite step of energy away. Think of a ladder. Attempting to step between the rungs would only bring you back to the lower one. You need to climb a definite height to reach the upper level. So it is with atoms. There are only certain allowed levels.

Max Karl Ernst Planck, 1858–1947, is well known for his amazing discovery of quantum theory and was awarded the Nobel Prize in Physics in 1918. How fortunate for science that he decided not to pursue his talent as a brilliant pianist instead. Less often mentioned are the tragedies in his life: three children died young, one of his two remaining sons was executed, his first wife died in 1909, and the hardship he endured in the last weeks of the war after his house was bombed.

This model is useful in understanding what happens during interpretation of spectra. However, the actual structure of the atom and behavior of electrons and other particles at the subatomic scale was found to be very different from the Newtonian macro-scale. Max Planck discovered that probabilities more appropriately described what happens at this dimension and proposed the quantum theory, in 1900, where energy only exists in packets, i.e., quantized. Niels Bohr, the Danish physicist, in 1913 used the quantum theory to refine and define energy transitions within an atom to lead to an understanding of spectral lines.

References are provided in the Appendix of this book for further study of the structure and properties of the atom. But for now it is necessary just to be aware of their existence while using simple models to delve into spectroscopy.

Now that you know a little about the interpretation of color and spectra, we can address the issue of classification of stars, which we will do in the next chapter.

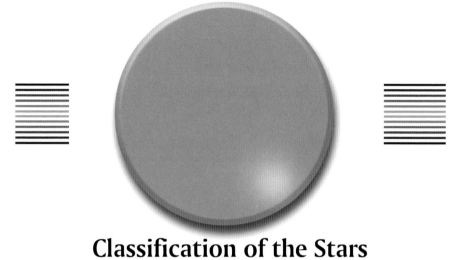

CHAPTER FOUR

Classification of the Stars

T. Buick, *The Rainbow Sky,* Patrick Moore's Practical Astronomy Series,
DOI 10.1007/978-1-4419-1053-0_4, © Springer Science + Business Media, LLC 2010

The Hertzsprung–Russell Diagram

Like Hipparchus, we can look up into the sky and make judgments or guesses about the size and brightness of the twinkling marvels. The winter constellation of the Pleiades in the northern hemisphere's sky, shown in Fig. 4.1, illustrates the varying degrees of brightness that we recognize. Modern astronomy has expanded on this knowledge, as we have seen, so much so that some properties of a huge number of stars can be graphically represented to enable comparisons at a glance. One such plot was constructed by the Danish astronomer Ejnar Hertzsprung in 1906 for stars in the clusters of the Pleiades, or M45, and the Hyades, in the constellation of Taurus. Soon after, Henry Russell, an American astronomer, independently carried out a similar exercise, and the plot is now known as the Hertzsprung–Russell, or H–R, diagram. The H–R diagram consists of a plot of the brightness of stars, their luminosity or absolute magnitude, as a function of their temperature or spectral class (*see* later). There are many representations of the diagram, and Fig. 4.2 is a particularly

Fig. 4.1. The brightness of the stars varies within the familiar pattern of the Pleiades.

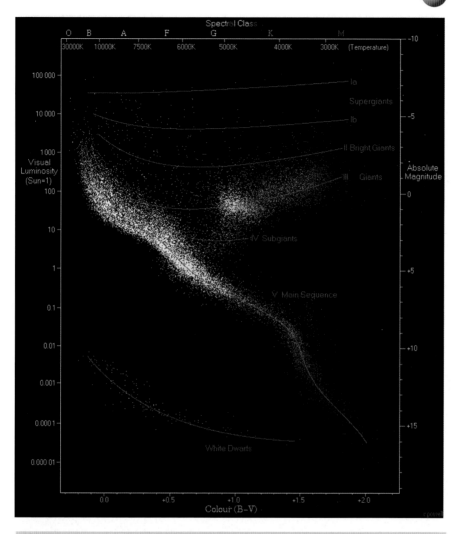

Fig. 4.2. The Hertzsprung–Russell diagram (Credit: Richard Powell).

colorful and informative representation of 22,000 stars from the Hipparcos catalog and 1,000 red and white dwarfs from the Gliese *Catalogue of Nearby Stars.*

Ejnar Hertzsprung (1873–1967) was born in Copenhagen and became director of an insurance company. He involved himself with the technology of acetylene lighting, stereo photography, then spectrophotometry before pursuing astronomy at the Copenhagen University. His tutor once said of him, "I think occasionally. Ejnar thinks all the time."

It is particularly gratifying that there are so many amateur astronomy societies these days making an important contribution to the science of the night sky. Within those societies are members with much expertise and dedication; an example is the following creation of an HR diagram constructed from 150 stars actually seen during observing sessions throughout the year (*see* Fig. 4.3).

The diagram is accompanied by notes, tables, and properties of the stars. Concerning the diagram, Harries writes:

A Hertzsprung–Russell diagram plots absolute magnitude against spectral type. In order to produce a list of stars for an HR diagram that is both representative and meaningful to the observer,

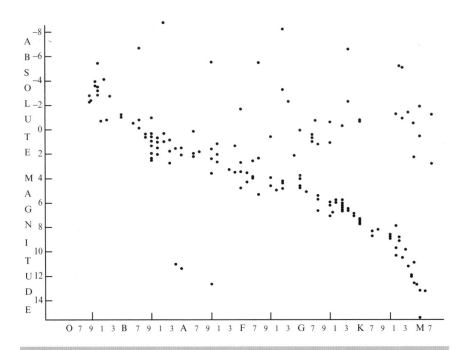

Fig. 4.3. The brightness and class of around 150 stars (Credit: Jeff Harries of the Orpington Astronomical Society, UK).

certain criteria have to be borne in mind. A chart containing only the brightest stars in the sky would be dominated by those that are intrinsically luminous, which would give a clumping of examples at the top end of the scale. Conversely, using only the nearest stars would have the opposite effect of producing a chart that was dominated by examples of low luminosity at the bottom. I have therefore tried to steer a middle way by choosing only stars that have been observed, which drastically reduces the number of our closest stellar neighbors, while offsetting this by giving these less luminous stars observational priority.

I thought it necessary to have all first magnitude stars on the list, together with true stellar companions, so have included half a dozen stars that have not actually been seen, together with the 150 that have. Because our local skies are often not conducive to viewing deep sky objects, doubles and multiple stars form a more reliable foundation around which to plan a night's observing. Differing offsets on the chart indicate that some of these doubles and multiples are only line of sight companions. Special interest stars have been selected for their spectral features, to give a better overall representation. There are a number of variables, and three are runaway stars from the star-forming nursery of the Orion Nebula: Mu Columbæ, AE Aurigæ, and 53 Arietis.

Those of us at Orpington Astronomical Society, England, who have been to the southern hemisphere have seen the various stars on the chart with low declinations visually, while those who have visited the Canary Isles had the optical aid of binoculars or better. Our society visits to COAA in Portugal, during 2007, allowed us the use of their 20" reflector to observe some testing objects, while nearer to home our Deep Sky camps have given us the benefit of darker skies at a variety of sites in deepest Kent. The majority of our observing is still done in and around the Orpington area, especially at High Elms Country Park, where we hold our regular Observing Evenings.

Here is a key for the list of 150 observed stars used in the HR diagram:

Con	Constellation
RA	Right Ascension (in hours, minutes, seconds)
Dec	Declination (in degrees, minutes, seconds)
Mag	Apparent magnitude taken from the Hipparcos catalog
Offset	Difference between apparent and absolute magnitude – calculated from the parallax given in the Hipparcos catalog
Ab Mag	Absolute magnitude (Magnitude + Offset = Intrinsic luminosity)

Separation	Magnitude of components and separation (seconds of an arc)
n/s	Not seen
Spectral type	O/B,the hottest and bluest stars, through to M, the coolest and reddest (for the purposes of the chart, C and S stars are taken as being the same as M, unless described as otherwise in their spectral type e.g. S7 plotted as if it is M7)

Code:

1*	First magnitude stars
D	Double/Multiple stars
SI	Special interest stars
VC	Cepheid variable
VI	Irregular variable
VM	Mira variable
VR	RR Lyra variable
VS	Semi-regular variable

Scope: 7×50s and 10×50s are binoculars, 2½ , 4, and 5 are refracting telescopes; 12 LB for Lightbridge and 20 are reflecting telescopes (Table 4.1).

The brightness that we see as we peer at a star in the night sky is, quite sensibly, termed the apparent magnitude. However, as the seasons pass we may see that star close to the horizon, and the light passes through a thicker layer of Earth's atmospheric gases and dust than it did when observed overhead. A slight loss of brightness occurs because of light scattering, and since shorter wavelengths, or bluer light, is scattered more than longer or redder wavelengths a reddening is observed compared with the overhead vision. For the same reason reddening occurs as radiation (light) travels through the interstellar medium of space. Apparent magnitude is therefore not the best parameter to work with.

> The space between the stars, or interstellar medium, is not empty but contains ions, atoms, molecules, and dust particles as well as magnetic fields and cosmic rays. Much of the gas is hydrogen and helium, but much larger molecules and familiar substances are there.

Table 4.1. Data for 150 stars observed.

Name	Con	RA h	RA m	RA s	Dec °	Dec ′	Dec ″	Mag	Offset	Ab Mag	Separation	Spectral Type	Code	Scope	Location
Groombridge 34A	And	0	18	20.54	44	1	19.0	8.09	2.24	10.33	8.09+11.04 @35″	M1V	D	5″	Orpington
Groombridge 34B								11.04	2.24	13.28		M6V			″
96 G. Piscium	Psc	0	48	22.53	5	17	0.2	5.74	0.64	6.38		K2V	SI	10×50s	Tuesnoad
η Cas A	Cas	0	49	5.10	57	48	59.6	3.46	1.13	4.59	3.46+7.53 @13″	G0V	D	2½″	Orpington
η Cas B								7.53	1.13	8.66		K7V			″
ψ¹ Psc A	Psc	1	5	40.93	21	28	23.6	5.33	-4.32	1.01	5.33+5.55 @30.0″	A1vn	D	10×50s	Orpington
ψ¹ Psc B		1	5	41.68	21	27	55.7	5.55	-4.17	1.38		A0vn			″
ζ Psc A	Psc	1	13	43.80	7	34	31.8	5.21	-3.28	1.93	5.21+6.44 @23.0″	A7IV	D	2½″	Tuesnoad
ζ Psc B		1	13	45.17	7	34	42.2	6.44	-3.89	2.55		F7V			″
HD 8357	Psc	1	22	56.70	7	25	7.3	7.30	-3.28	4.02		G5+X	SI	10×50s	Tuesnoad
α Eri	Eri	1	37	42.75	-57	14	12.0	0.45	-3.22	-2.77		B3Vp	1*	Visual	Australia
τ Cet	Cet	1	44	5.13	-15	56	22.4	3.49	2.19	5.68		G8V	SI	Visual	Orpington
γ Ari A	Ari	1	53	31.77	19	17	38.7	4.66	-3.98	0.68	4.66+4.60 @7.5″	A1pSi	D	2½″	Orpington

(continued)

Table 4.1. (continued).

Name	Con	RA h	RA m	RA s	Dec °	Dec '	Dec "	Mag	Offset	Ab Mag	Separation	Spectral Type	Code	Scope	Location
γ Ari B								4.60	−3.98	0.62		B9V			"
66 Cet A	Cet	2	12	47.32	−2	23	36.5	5.65	−3.32	2.33	5.65+ 7.56 @ 16.5"	F8V	D	2½"	Orpington
66 Cet B		2	12	46.43	−2	23	46.2	7.56	−5.47	2.09		G4			"
268 G. Ceti	Cet	2	36	3.83	6	53	0.1	5.79	0.71	6.50		K3V	SI	10×50s	Orpington
θ Per A	Per	2	44	11.64	49	13	43.2	4.10	−0.25	3.85	4.10+ 9.87 @ 20.1"	F7V	D	12" LB	Tuesnoad
θ Per B								9.87	−0.25	9.62		M1V			"
53 Ari	Ari	3	7	25.69	17	52	49.9	6.13	−6.82	−0.69		B1.5V	SI	7×50s	Orpington
82 G. Eridani	Eri	3	19	55.70	−43	4	11.0	4.26	1.09	5.35		G8V	SI	Visual	Tenerife
ε Eri	Eri	3	32	56.48	−9	27	29.9	3.72	2.46	6.18		K2V	SI	10×50s	Orpington
X Per	Per	3	55	23.08	31	2	45.1	6.79	−9.59	−2.80		O9.5pe+ X	SI	7×50s	Orpington
o² Eri A	Eri	4	15	16.30	−7	39	10.0	4.43	1.49	5.92	4.43+9.31 @ 83"	K1V	D	5"	Orpington
o² Eri B								9.52	1.49	11.01	9.52+11.19 @ 9"	A4VII			"
o² Eri C								11.19	1.49	12.68		M4.5V			"

θ¹ Tau	Tau	4	28	34.43	15	57	44.0	3.84	-3.42	0.42	3.84+3.40 @337"	G7III	D	Visual	Headcorn
θ² Tau		4	28	39.67	15	52	15.4	3.40	-3.30	0.10		A7III			"
Aldebaran	Tau	4	35	55.20	16	30	35.1	0.87	-1.50	-0.63		K5III	1*	Visual	Headcorn
55 Eri A	Eri	4	43	34.73	-8	47	39.3	6.67	-5.47	1.20	6.67+6.79 @9.2	G8III	D	2½"	Orpington
55 Eri B								6.79	-5.47	1.32		F4IIIpSr			"
Rigel	Ori	5	14	32.27	-8	12	5.9	0.18	-6.87	-6.69		B8Ia	1*	Visual	Orpington
AE Aur	Aur	5	16	18.15	34	18	44.0	5.99	-8.25	-2.26		O9.5Vvar	SI	10×50s	Orpington
Capella A	Aur	5	16	41.30	45	59	56.5	0.61	-0.56	0.05	0.61+1.11 @0.05" n/s	G5III	1* D	Visual	Orpington
Capella B								1.11	-0.56	0.55		G0III			"
HD 36395	Ori	5	31	26.95	-3	40	19.7	7.97	1.22	9.19		M1.5V	SI	10×50s	Orpington
γ Lep A	Lep	5	44	27.97	-22	26	51.0	3.59	0.24	3.83	3.59+6.15 @97"	F7V	D	10×50s	Orpington
γ Lep B								6.15	0.24	6.39		K2V			
μ Col	Col	5	45	59.89	-32	18	23.0	5.16	-7.99	-2.83		B1IV/V	SI	7×50s	Tenerife
Betelgeuse	Ori	5	55	10.29	7	24	25.3	0.45	-5.59	-5.14		M2Ib	1*	Visual	Orpington
8 Mon A	Mon	6	23	46.10	4	35	34.2	4.39	-2.98	1.46	4.39+6.72 @13.4	A5IV	D	2½"	Headcorn
8 Mon B		6	23	46.50	4	35	45.1	6.72	-1.93	4.79		F5V		Visual	"
Canopus	Car	6	23	57.09	-52	41	44.6	-0.62	-4.91	-5.53		F0Ib	1*	Visual	Tenerife

(continued)

Table 4.1. (continued).

Name	Con	RA h	RA m	RA s	Dec °	Dec '	Dec "	Mag	Offset	Ab Mag	Separation	Spectral Type	Code	Scope	Location
Sirius A	CMa	6	45	9.25	−16	42	47.3	−1.44	2.89	1.45	1.44+8.49 @ 6" n/s	A1V	1* D	Visual	Orpington
Sirius B								8.49	2.89	11.38		wdA5			"
ε CMa	CMa	6	58	37.55	−28	58	19.5	1.50	−5.60	−4.10		B2II	SI	Visual	Orpington
l² Pup	Pup	7	13	32.25	−44	38	25.9	4.42	−3.92	0.50		M5e	VS	7×50s	Tenerife
h 3945 A	CMa	7	16	36.84	−23	18	56.2	4.83	−11.42	−6.59	4.83+6.02 @ 26.4"	K3Ib	D	2½"	Tuesnoad
h 3945 B		7	16	38.38	−23	18	39.8	6.02	−4.49	1.53		F0			"
Luyten's Star	CMi	7	27	24.16	5	14	5.2	9.84	2.10	11.94		M3.5V	SI	2½"	Orpington
n Pup A	Pup	7	34	18.67	−23	28	25.2	5.79	−2.32	3.47	5.79+5.83 @ 9.6"	F4V	D	2½"	Headcorn
n Pup B								5.83	−2.32	3.51		F6V			"
Castor A	Gem	7	34	36.00	31	53	19.1	1.94	−0.99	0.95	1.94+2.92 @ 4.4"	A2V	D	5"	Orpington
Castor B								2.92	−0.99	1.93		Am			"
YY Gem	Gem							8.80	−0.99	7.81	8.80 @ 72.5" to Castor	M1	D	5"	"
k Pup A	Pup	7	38	49.88	−26	48	14.0	4.46	−5.72	−1.26	4.46+4.66 @ 9.9	B5IV	D	2½"	Orpington
k Pup B								4.66	−5.72	−1.06		B5IV			"

Procyon A	CMi	7	39	18.54	5	13	39.0	0.40	2.28	2.68	0.40+10.35 @ 5" n/s	F5IV	1*	D	Visual	Orpington
Procyon B								10.35	2.28	12.63		dF				"
Pollux	Gem	7	45	19.36	28	1	34.7	1.16	-0.07	1.09		K0IIIvar	1*		Visual	Orpington
ι¹ Cnc A	Cnc	8	46	41.83	28	45	36.0	4.03	-4.80	-0.78	4.03+6.58 @ 30.5"	G7.5IIIaBa		D	2½"	Headcorn
ι¹ Cnc B		8	46	40.00	28	45	54.6	6.58	-3.81	2.76		A3V		D	2½"	"
Σ1321 A	UMa	9	14	24.27	52	41	16.8	7.64	1.04	8.68	7.64+7.70 @ 17.1"	M0		D	2½"	Orpington
Σ1321 B		9	14	26.19	52	41	16.7	7.70	1.01	8.71		M0				"
Regulus	Leo	10	8	22.46	11	58	1.9	1.36	-1.88	-0.52		B7V	1*		Visual	Orpington
Groombridge 1618	UMa	10	11	23.36	49	27	19.7	6.60	1.56	8.16		K8V		SI	10×50s	Orpington
γ Leo A	Leo	10	19	58.16	19	50	30.7	2.31	-2.93	-0.62	2.31+3.56 @ 4.4"	K0III		D	4"	Orpington
γ Leo B								3.56	-2.93	0.63		G7II				"
AD Leo	Leo							9.32	1.55	10.87	9.32 @ 6' to γ Leo	M4		D	4"	"
Lalande 21185	UMa	11	3	20.61	35	58	53.3	7.49	2.97	10.46		M2V		SI	7×50s	Orpington
Groombridge 1830	UMa	11	52	55.82	37	43	58.1	6.42	0.19	6.61		G8Vp		SI	10×50s	Orpington
2 CVn A	CVn	12	16	7.54	40	39	36.9	5.69	-7.04	-1.35	5.69+8.64 @ 11.4"	M1III		D	2½"	Orpington
2 CVn B		12	16	6.54	40	39	34.8	8.64	-3.35	5.29		F8V				"

(continued)

Table 4.1. (continued).

Name	Con	RA h	RA m	RA s	Dec °	Dec '	Dec "	Mag	Offset	Ab Mag	Separation	Spectral Type	Code	Scope	Location
α Cru A	Cru	12	26	35.94	-63	5	56.6	1.33	-4.96	-3.63	1.33+1.73 @4" n/s	B0.5IV	1* D	Visual	Australia
α Cru B								1.73	-4.96	-3.23		B1V			"
γ Cru	Cru	12	31	9.93	-57	6	45.2	1.59	-2.15	-0.56		M4III	SI	Visual	Australia
Y CVn	CVn	12	45	7.83	45	26	24.8	5.42	-6.69	-1.27		C7Iab	VS	10×50s	Orpington
β Cru	Cru	12	47	43.32	-59	41	19.4	1.25	-5.17	-3.92		B0.5III	1*	Visual	Australia
α CVn A	CVn	12	56	1.84	38	19	5.7	2.89	-2.64	0.25	2.89+5.61 @19.4"	A0pSiEuHg	D	2½"	Orpington
α CVn B		12	56	0.60	38	18	52.9	5.61	-2.03	3.58		F0V			"
61 Vir	Vir	13	18	24.97	-18	18	31.0	4.74	0.35	5.09		G6V	SI	10×50s	Orpington
Mizar A	UMa	13	23	55.42	54	55	31.5	2.23	-1.90	0.33	2.23+3.91 @14.4"	A2V	D	2½"	Orpington
Mizar B								3.91	-1.90	2.01		A1m			"
Spica	Vir	13	25	11.60	-11	9	40.5	0.98	-4.53	-3.55		B1V	1*	Visual	"
Alcor	UMa	13	25	13.42	54	59	16.8	3.99	-1.98	2.03	3.99 @708 to Mizar	A5V	D	Visual	Orpington
β Cen	Cen	14	3	49.40	-60	22	23.0	0.61	-6.03	-5.42		B1III	1*	Visual	Australia
Arcturus	Boö	14	15	40.35	19	11	14.2	-0.05	-0.26	-0.31		K2IIIp	1*	Visual	Orpington
α Centuri A	Cen	14	39	40.30	-60	50	6.5	-0.01	4.35	4.34	:-0.01+1.35 @10 n/s	G2V	1* D	Visual	Australia
α Centuri B		14	39	39.39	-60	50	22.1	1.35	4.35	5.70		K1V			"
Proxima Centauri								11.01	4.44	15.45	11.01 @131' to α n/s	dM5V			"

Name	Const	h	m	s	°	'	"	mag			Double	Spectrum		Aperture	Observatory
α¹ Lib	Lib	14	50	52.78	-16	2	29.8	2.75	-1.87	0.88	2.75+5.15 @231"	A3IV	D	10×50s	Woodlands
α² Lib	Lib	14	50	41.26	-15	59	49.5	5.15	-1.87	3.28		F3V	D		"
33 G. Librae A	Lib	14	57	27.35	-21	24	40.6	5.72	1.14	6.86	5.72+8.01 @25"	K4V	D	2½"	Woodlands
33 G. Librae B								8.01	0.63	8.64		M0			"
Washington 5583	Lib	15	10	13.69	-16	22	14.9	9.07	-2.33	6.74	9.07+9.44 @6'	K0/K1V	D	2½"	Orpington
Washington 5584	Lib	15	10	13.58	-16	27	15.5	9.44	-2.36	7.08		K0V			"
β Lib	Lib	15	17	0.47	-9	22	58.3	2.61	-3.45	-0.84		B8V	SI	Visual	Woodlands
R CrB	CrB	15	48	34.42	28	9	24.4	5.89	-11.34	-5.45		C0,0(F8pe)	VI	7×50s	Headcorn
σ CrB A	CrB	16	14	41.70	33	51	35.0	5.58	-1.68	3.90	5.58+6.59 @7.2"	G0VCalle	D	2½"	Orpington
σ CrB B	CrB	16	14	42.00	33	51	35.0	6.59	-1.68	4.91		G1V	D	7×50s	"
v¹ CrB	CrB	16	22	21.40	33	47	57.0	5.20	-6.16	-0.96	5.20+5.40 @6'	M2III	D	7×50s	Orpington
v² CrB	CrB	16	22	29.22	33	42	12.1	5.40	-6.12	-0.73		K5III			"
Antares A	Sco	16	29	24.47	-26	25	55.0	1.06	-6.33	-5.28	1.06+5.50 @2.5"	M1.5Iab	1*D	20"	COAA
Antares B								5.50	-6.33	-0.83		B2.5V			"
36 Oph A	Oph	17	15	21.29	-26	36	0.2	5.07	1.11	6.18	5.07+5.10 @5.0"	K0V	D	20"	COAA

(continued)

Table 4.1. (continued).

Name	Con	RA h	RA m	RA s	Dec °	Dec'	Dec"	Mag	Offset	Ab Mag	Separation	Spectral Type	Code	Scope	Location
36 Oph B								5.10	1.11	6.21		K2V			"
36 Oph C		17	16	13.68	−26	32	36.3	6.33	1.12	7.45	6.33 @ 12.2' to A/B	K5V			"
v¹ Dra	Dra	17	32	10.42	55	11	2.8	4.89	−2.41	2.48	4.89+4.86 @ 62"	Am	D	7×50s	Orpington
v² Dra		17	32	15.88	55	10	22.1	4.86	−2.43	2.43		Am			"
Barnard's star	Oph	17	57	50.00	4	42	0.0	9.54	3.70	13.24		M5	SI	2½"	Orpington
98 Her	Her	18	6	1.91	22	13	8.0	4.96	−6.34	−1.42		M3IIIaBa 0.2ZrO	SI	10×50s	Orpington
Vega Lyr	lyr	18	36	56.19	38	46	58.8	0.03	0.55	0.58		A0Vvar	1*	Visual	Orpington
Σ2398 A	Dra	18	42	48.22	59	37	33.7	8.94	2.24	11.18	8.94+9.70 @ 15.3"	M3V	D	5"	Orpington
Σ2398 B		18	42	48.51	59	37	20.5	9.70	2.27	11.97		M3.5V			
ε¹ lyr A	lyr	18	44	20.34	39	40	11.9	5.00	−3.48	1.52	4.66+4.59 @ 208"	A4V	D	Visual	Orpington
ε¹ lyr B								6.10	−3.48	2.62	5.00+6.10 @ 2.4"	F1V	D	5"	"
ε² lyr A		18	44	22.78	39	36	45.3	5.23	−3.46	1.77	5.23+5.47 @ 2.4"	A8Vn	D	5"	"

Star	Con											Spectral			Location
ε² Lyr B								5.47	-3.46	2.01		F1Vn			"
ζ Lyr A	Lyr	18	44	46.34	37	36	18.2	4.34	-3.37	0.97	4.34+5.73 @43.7"	Am	D	10×50s	Orpington
ζ Lyr B		18	44	48.19	37	35	40.4	5.73	-3.32	2.41		F0IVvar			"
κ Pav	Pav	18	56	57.04	-67	14	0.7	4.40	-6.11	-1.71		F5Ib-II	VC	Visual	Australia
RR Lyr	Lyr	19	25	28.00	42	47	5.4	7.92	-6.79	1.13		A5.0-F7.0	VR	7×50s	Orpington
β Cyg A	Cyg	19	30	43.29	27	57	34.9	3.05	-5.36	-2.31	3.05+5.12 @34.4"	K3II+···	D	2½"	Orpington
β Cyg B		19	30	45.40	27	57	55.0	5.12	-5.31	-0.19		B8V	SI	Visual	Orpington
σ Dra	Dra	19	32	20.59	69	39	55.4	4.67	1.20	5.87		K0V	SI	Visual	Orpington
16 Cyg A	Cyg	19	41	49.09	50	31	31.6	5.99	-1.67	4.32	5.99+6.25 @40"	G2V	D	10×50s	Orpington
16 Cyg B		19	41	52.10	50	31	31.6	6.25	-1.65	4.60		G5V			"
17 Cyg A	Cyg	19	46	25.58	33	43	43.3	5.00	-1.60	3.40	5.00+9.20 @26.0"	F5V	D	2½"	Orpington
17 Cyg B								9.20	-1.60	7.60		K5V			"
Chi Cyg	Cyg	19	50	33.94	32	54	50.9	7.91	-5.13	2.78		S7,1e	VM	Visual	Tuesnoad
Altair	Aql	19	50	46.68	8	52	2.6	0.76	1.44	2.20		A7IV-V	1*	Visual	Orpington
HDE 226868	Cyg	19	58	24.00	35	12	0.0	8.84	-11.18	-2.34		O9.7Iab+X	SI	2½"	Orpington
δ Pav	Pav	20	8	41.86	-66	10	45.6	3.55	1.07	4.62		G5IV-Vvar	SI	Visual	Australia
HD 191408 A	Sgr	20	11	11.61	-36	5	50.6	5.32	1.09	6.41	5.32+11.50 @7"	K2V	D	20"	COAA

(continued)

Table 4.1. (continued).

Name	Con	RA h	RA m	RA s	Dec °	Dec '	Dec "	Mag	Offset	Ab Mag	Separation	Spectral Type	Code	Scope	Location
HD 191408 B								11.50	1.09	12.59		M4V			"
α¹ Cap	Cap	20	17	38.86	-12	30	29.6	4.30	-6.62	-2.32	4.30+3.58 @ 378"	G3Ib	D	Visual	Headcorn
α² Cap		20	17	39.00	-12	30	30.0	3.58	-2.61	0.97		G6/G8III	1*	Visual	"
Deneb	Cyg	20	41	25.91	45	16	49.2	1.25	-9.98	-8.73		A2Ia	1*	Visual	Orpington
61 Cyg A	Cyg	21	6	50.24	38	44	29.0	5.20	2.29	7.49	5.20+6.05 @ 30.6"	K5V	D	10×50s	Orpington
61 Cyg B		21	6	52.19	38	44	3.9	6.05	2.28	8.33		K7V			"
Lacaille 8760	Mic	21	17	17.71	-38	51	52.5	6.69	2.02	8.71		M1/M2V	SI	7×50s	Tenerife
γ Pav	Pav	21	26	26.49	-65	22	5.3	4.21	0.18	4.39		F6V	SI	Visual	Australia
ε Ind	Ind	22	3	17.44	-56	46	47.3	4.69	2.20	6.89		K5V	SI	Visual	Australia
δ Cep A	Cep	22	29	10.25	58	24	54.7	4.07	-7.39	-3.32	4.07+6.31 @ 41.0"	G2Ibvar	VC D	2½"	Orpington
δ Cep B		22	29	9.23	58	24	14.7	6.31	-7.32	-1.01		A0			"
TW PsA	PsA	22	56	23.83	-31	33	54.6	6.48	0.59	7.07		K4Vp	SI	7×50s	Tenerife
Fomalhaut	PsA	22	57	38.83	-29	37	18.6	1.17	0.57	1.74		A3V	1*	Visual	Orpington
Lacaille 9352	PsA	23	5	47.17	-35	51	22.7	7.35	2.41	9.76		M2/M3V	SI	7×50s	Tenerife

Bradley 3077	Cas	23	13	14.74	57	18	3.5	5.57	0.93	6.50	K3Vvar	SI	10×50s	Orpington
ET And	And	23	17	55.99	45	29	20.2	6.48	-6.12	0.36	B9p…	SI	10×50s	Orpington
94 Aqr A	Aqr	23	19	6.37	-13	27	17.0	5.30	-1.58	3.72	G5IV	D	2½"	Tuesnoad
94 Aqr B								7.30	-1.58	5.72	K2V		"	"
R Aqr	Aqr	23	43	49.44	-15	17	3.9	8.69	-6.47	2.22	M4pe	SI	2½"	Orpington
19 Psc	Psc	23	46	23.54	3	29	12.7	4.95	-6.84	-1.89	C5II	SI	10×50s	Orpington
ρ Cas	Cas	23	54	23.00	57	29	58.0	4.51	-12.76	-8.25	G2 0e	VI	Visual	Orpington
Sun		24	0	0.00	0	0	0.0	-26.74	31.57	4.83	G2V	SI	Visual	Orpington

94 Aqr A: 5.30+7.30 @ 12.7"

The standardized or absolute magnitude with which to compare stars is the intrinsic brightness after allowing for distance, the standard distance for magnitude correction being 10 parsecs. (One parsec is equal to 3.2616 light years or 30.857×10^{12} km.) The position of a star also matters because there is more dust in the plane of our galaxy and a variation in density between galactic arms and known molecular clouds. The effect of dimming through space is known as extinction. Our Sun has an apparent magnitude from Earth of a huge -26.74 but an absolute magnitude of just $+4.83$, brightness hardly visible in the average light-polluted night sky.

Star Colors

The color of a star is directly related to its temperature, so it should be easy to look at a star, estimate its color, and enter it into a graph along with the magnitude to construct an H–R diagram. Unfortunately, there is a catch. The eye, and its perception of color, varies from human to human, so each of us might place the dot in a different place on the graph. How does the eye work, and what happens when we look at small dots of light?

The eye is an extremely sensitive detector of photons; only a few are required to initiate a response in our eyes. It might be predicted then that we should be able to perceive the colors of stars at very low light levels. But we do not! The sensory retina layer at the back of the eye is stuffed with millions of photoreceptors of two types, rods and cones. The 7 million cones are of three types – red, blue, and green – and are clustered more towards the center of the retina. The 120 million rods are far more sensitive and spread predominantly away from the center. However, there is a price to pay for being more sensitive; the rods can only see images in monochrome or gray scale, not full color, which is why we cannot see color in the dimness of the night. This is known as the Purkinje effect. Many a disbeliever has taken photographs under bright moonlight to illustrate the difference between the actual color of flowers and what the eye sees.

> Here's an experiment to simulate the Purkinje effect (by a different mechanism). Try to note the difference between your perception of color intensity when facing red and green objects together or turning your back on them, bending down, and viewing from between your legs!

As if that were not enough, the eye is very poor at distinguishing colors in point sources, a condition known as small field tritanopia (a sort of color blindness). A later chapter will show how to get around this problem.

To remove error-prone subjective color estimation, spectroscopes or spectrographs are used to detect and record the intensity of light at different wavelengths. A manageable color parameter – spectral class – is now available for use in the H–R diagram. Remember that temperature and color are directly related.

The idea of spectral class as a representation of temperature (and composition), and therefore color, evolved over a long period of time. Optics had its origins in Greece, especially in the efforts of Euclid (330–260 B.C.), who stated many of the then-known laws of light, such as the angle of incidence is equal to the angle of reflection. Colors and light were certainly concepts thought about by scientists and philosophers through the centuries.

Roger Bacon in the thirteenth century was interested in the optics of rainbows. It was held then that colors arose from the modification of light and that light appears white in its pristine form. Newton disagreed with the idea of "modification" and asserted that light was a mixture that could be separated into its component parts – its colors. His experiments with prisms confirmed his theory and laid the foundations for much of what was to come in the development of the science of star colors, spectroscopy.

Personal anecdotes of ancient eras abound, revealing that the early years of mathematics were dangerous times. Pythagoras (around 500 b.c.) is sometimes credited with the discovery that the angles of a triangle always add up to 180°, but Euclid was probably the first. One member of the Pythagorean secret society, according to legend, who divulged a property of triangles to his friends, fled but was tracked down to his vineyard and boiled alive in olive oil in front of his family. Also, according to legend, at that time Pythagoreans drowned Hippasus of Metapontum for heresy for daring to prove that some numbers could not be written as a fraction of integers!

Gustav Kirchhoff, with the chemist Robert Bunsen (of Bunsen burner fame) laid down the principles of spectral analysis that catalyzed the development of astronomical spectroscopy. Angelo Secchi, an Italian astronomer and priest, and William Huggins, an English amateur astronomer, in the 1860s categorized stars according to the properties of their spectra. Following a spectroscopic survey of 4,000 stars published in 1863, Secchi drew up a spectral classification of stars that was subsequently superseded by the Harvard system introduced by Edward Charles Pickering (American astronomer) in 1890, assisted by three women – Annie Jump Cannon, Williamina Paton Fleming, and Antonia Caetana de Paiva Pereira Maury. The modern system used is that created by William Morgan and Philip Keenan (MK system) and described in *An Atlas of Stellar Spectra* in 1943.

Kirchhoff's Laws, 1859

A solid, or gas under high pressure, when heated to incandescence, produces a continuous spectrum.

A gas under low pressure, but at a sufficiently high temperature, produces a spectrum of bright emission lines.

> A gas at low pressure and low temperature, lying between a hot continuum source and the observer, produces an absorption line spectrum, i.e., a number of dark lines superimposed on a continuous spectrum.

As stated, the color of a star is an important property and indicates temperature when constructing the H–R diagram. To be more precise, it is the color of the strongest or dominant color in its radiated spectrum. The intensity of emitted light (energy) varies with wavelength such that a hot star emits relatively more energy at blue wavelengths than red, while a cool star's emission peaks at orange or red. The intensities are expressed as apparent magnitudes.

So is the color of the most intense wavelength emitted recorded and used? Well, not quite. Monitoring a single wavelength can lead to substantial inaccuracies. To gain more accuracy, astronomers measure specific wavelengths and compare the two. Standard optical filters are now available. For instance, the Johnson filters U, B, V, R, and I allow light transmissions in the ultraviolet, blue, yellow-green, red, and near infrared. The system of measuring apparent magnitudes at two different wavebands forms the basis for defining the colors of stars and is called the color index, CI. The CI is simply the difference between the apparent magnitudes of two selected wavelengths. A common CI used is B–V. A hot, blue star around 15,000 K will emit more energy than a redder star of around 3,000 K. B–V for each will be their color indices. In other words, a star's color is given by its blue (B) magnitude minus its visible or green (V) magnitude. Because magnitudes are smaller for larger brightness, a brighter blue star will have a more negative Color Index. V–R is also used, although the B–V relationship will be most commonly encountered for reference to color (Fig. 4.4).

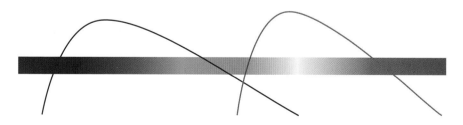

Fig. 4.4 This represents a very simplified illustration of the energy profile of a hot blue star and a cooler red star.

Table 4.2. (B–V) Color indices and temperature.

Star	(B–V) index	Temperature °K	Solar radii	Magnitude
Aldebaran	1.54	3,406 orange	219	0.84
Alnitak	−0.20	17,038 blue	17	1.71
Altair	0.22	7,758 blue	2.3	0.75
Aldebaran	1.54	3,406 orange	219	0.84
Arcturus	1.24	4,106 yellow	69	−0.07
Aldebaran	1.54	3,406 orange	219	0.84
Deneb	0.09	8,410 blue	297	1.25
Dubhe	1.06	4,589 yellow	59	1.78
Mizar	0.06	8,596 blue	4.4	2.21
Polaris	0.64	5,986 blue	71	1.96
Pollux	0.99	4,794 yellow	18	1.15
Procyon	0.43	6,800 blue	2.7	0.37
Rigel	−0.03	9,076 blue	99	0.15
Sirius	0.01	8,858 blue	2.4	−1.47
Vega	0.00	8,913 blue	3.6	0.03

It is often asked, "Since stars exist over a wide range of temperatures and colors why do we see lots of blue and red ones and not green ones?" Green is right in the middle of the human eye's response to light. When the star's emission maximum is green it is radiating strongly at all visible wavelengths and therefore appears as a mixture of those colors, that is, white.

Table 4.2 shows a selection of stars with their (B–V) color indices along with their approximate temperatures and VERY approximate visual colors. The size and magnitude of the stars is added for information but require other relevant parameters to indicate their relationship with the color indices. Arranged in the stars' alphabetical order the colors and indices are random.

Through the magic of computing, arranging the (B–V) column in ascending order, the fill colors and temperature values can be seen to correlate (*see* Table 4.3).

Although it is not possible to present an exact visual of the color relationship with the (B–V) index Table 4.4 helps to give a better idea of colors actually observed by the "average" eye. Mitchell Charity has also converted color indices to pixel rgb values in the table. For complete

Table 4.3. Rearranged (B–V) color indices and temperature.

Star	(B–V) index	Temperature K	Solar radii	Magnitude
Alnitak	−0.20	17,038 blue	17	1.71
Rigel	−0.03	9,076 blue	99	0.15
Vega	0.00	8,913 blue	3.6	0.03
Sirius	0.01	8,858 blue	2.4	−1.47
Mizar	0.06	8,596 blue	4.4	2.21
Deneb	0.09	8,410 blue	297	1.25
Altair	0.22	7,758 blue	2.3	0.75
Procyon	0.43	6,800 blue	2.7	0.37
Polaris	0.64	5,986 blue	71	1.96
Pollux	0.99	4,794 yellow	18	1.15
Dubhe	1.06	4,589 yellow	59	1.78
Arcturus	1.24	4,106 yellow	69	−0.07
Betelgeuse	1.50	3,488 orange	1516	0.43
Aldebaran	1.54	3,406 orange	219	0.84
Antares	1.86	2,776 orange	7329	1.03

caveats and explanations visit his website – given in the Appendix of this book.

Since the H–R diagram was first put together a century ago some wonderful graphics have been drawn for a multitude of stars. Remember that the luminosity or magnitude is plotted against surface temperature or (B–V) color index. There are many representations of the diagram. Refer back to Fig. 4.2, for a super representation of 22,000 stars from the Hipparcos catalog and 1,000 red and white dwarfs from the Gliese *Catalogue of Nearby Stars*.

The spectral class, color index, and temperature are closely related, as are magnitude and luminosity. Also, as already mentioned, the mass and age of a star are crucial parameters. Perhaps a little more should be said about spectral class, which helps us to imagine what each star looks like and is composed of.

Although many labels were put forward throughout history for star types, such as Roman numerals and sequential letters of the alphabet, the ones used now emanate essentially from those suggested by Annie Jump Cannon in 1901, OBAFGKM. (A favorite mnemonic to remember

Table 4.4. Color for (B–V) Indices.

B-V	T_{eff}		B-V	T_{eff}		B-V	T_{eff}		B-V	T_{eff}	
-0.40	113017	#9bb2ff	0.25	7483	#eeefff	0.90	5052	#ffe8ce	1.55	3892	#ffd29c
-0.35	56701	#9eb5ff	0.30	7218	#f3f2ff	0.95	4948	#ffe6ca	1.60	3779	#ffd096
-0.30	33605	#a3b9ff	0.35	6967	#f8f6ff	1.00	4849	#ffe5c6	1.65	3640	#ffcc8f
-0.25	22695	#aabfff	0.40	6728	#fef9ff	1.05	4755	#ffe3c3	1.70	3463	#ffc885
-0.20	16954	#b2c5ff	0.45	6500	#fff9fb	1.10	4664	#ffe2bf	1.75	3234	#ffc178
-0.15	13674	#bbccff	0.50	6285	#fff7f5	1.15	4576	#ffe0bb	1.80	2942	#ffb765
-0.10	11677	#c4d2ff	0.55	6082	#fff5ef	1.20	4489	#ffdfb8	1.85	2579	#ffa94b
-0.05	10395	#ccd8ff	0.60	5895	#fff3ea	1.25	4405	#ffddb4	1.90	2150	#ff9523
-0.00	9531	#d3ddff	0.65	5722	#fff1e5	1.30	4322	#ffdbb0	1.95	1675	#ff7b00
0.05	8917	#dae2ff	0.70	5563	#fffefe0	1.35	4241	#ffdaad	2.00	1195	#ff5200
0.10	8455	#dfe5ff	0.75	5418	#ffeddb	1.40	4159	#ffd8a9			
0.15	8084	#e4e9ff	0.80	5286	#ffebd6	1.45	4076	#ffd6a5			
0.20	7767	#e9ecff	0.85	5164	#ffe9d2	1.50	3989	#ffd5a1			

Table courtesy of Mitchell Charity

the sequence is "Oh be a fine girl kiss me.") Cannon then added numbers to denote the relative positions between the letters. The nomenclature has since been slightly refined to indicate chemical composition obtained from spectral information. Table 4.5 presents a list of the major star classes and an approximate idea of the colors. A better representation is given in Tables 4.6 and 4.7 for the various spectral classes (courtesy of Mitchell Charity). A reference to a source of fully comprehensive tables is given in the Appendix of this book.

Table 4.5. Star class properties.

Spectral class	Temperature °K	Color
O	50,000–28,000	Blue
B	28,000–10,000	Blue-white
A	10,000–7,500	White
F	7,500–6,000	White-yellow
G	6,000–4,900	Yellow
K	4,900–3,500	Orange
M	3,500–2,000	Red

Table 4.6. Stellar types.

	r	g	b	rgb
O5(V)	157	180	255	#9db4ff
B1(V)	162	185	255	#a2b9ff
B3(V)	167	188	255	#a7bcff
B5(V)	170	191	255	#aabfff
B8(V)	175	195	255	#afc3ff
A1(V)	186	204	255	#baccff
A3(V)	192	209	255	#c0d1ff
A5(V)	202	216	255	#cad8ff
F0(V)	228	232	255	#e4e8ff
F2(V)	237	238	255	#edeeff
F5(V)	251	248	255	#fbf8ff
F8(V)	255	249	249	#fff9f9
G2(V)	255	245	236	#fff5ec
G5(V)	255	244	232	#fff4e8
G8(V)	255	241	223	#fff1df
K0(V)	255	235	209	#ffebd1
K4(V)	255	215	174	#ffd7ae
K7(V)	255	198	144	#ffc690
M2(V)	255	190	127	#ffbe7f
M4(V)	255	187	123	#ffbb7b
M6(V)	255	187	123	#ffbb7b

Table 4.7. Some nearby stars.

Sun (G2V) #fff5f2	**Sirius B** (DA2) #a8bdff *(as D?2)*	**EZ Aquarii** (M5.0V) #ffcc6f	**epsilon Indi** (K5Ve) #ffd2a1 *(as K5V)*
Alpha Centauri A (G2V) #fff5f2	**UV Ceti A** (M5.5V) #ffcc6f *(as M5V)*	**Procyon** (F5IV-V) #f1efff *(as F5IV)*	**DX Cancri** (M6.5V) #ffc370 *(as M6V)*
Alpha Centauri B (K0V) #ffeedd	**UV Ceti B** (M6.0V) #ffc370	**61 Cygni** (K5.0V) #ffd2a1	**tau Ceti** (G8Vp) #ffedde *(as G8V)*
Proxima Centauri (M5.5V) #ffcc6f *(as M5V)*	**Ross 154** (M3.5V) #ffce81 *(as M3V)* **Ross 248** (M5.5V) #ffcc6f *(as M5V)*	**61 Cygni B (?)** (K7.0V) #ffc78e	**RECONS 1** (M5.5V) #ffcc6f *(as M5V)*
Barnard's Star (M4.0V) #ffc97f	**epsilon Eridani** (K2V) #ffe3c4	**Gl 725 A** (M3.0V) #ffce81	**YZ Ceti** (M4.5V) #ffc97f *(as M4V)*
Wolf 359 (M6.0V) #ffc370		**Gl 725 B** (M3.5V) #ffce81 *(as M3V)*	**Luyten's Star** (M3.5V) #ffce81 *(as M3V)*
Lalande 21185 (M2.0V) #ffc483	**Lacaille 9352** (M1.5V) #ffcc8e *(as M1V)*	**Gl 15 A** (M1.5V) #ffcc8e *(as M1V)*	**Kapteyn's Star** (M1.5V) #ffcc8e *(as M1V)*
Sirius (A1V) #b5c7ff	**Ross 128** (M4.0V) #ffc97f	**Gl 15 B** (M3.5V) #ffce81 *(as M3V)*	

Star Composition

The importance of color and the spectroscopy of stars cannot be over-emphasized. The amount of information within their spectra is huge, which is just as well since there is currently no other way of analyzing stars with the possible exception of our own star, the Sun. The apparent change in frequency and wavelength of a wave, the shift, is called the Doppler effect and is due to the movement of the source of the wave relative to the observer, much like the sound varying as an emergency vehicle siren speeds toward or away from you.

> The Genesis mission launched a spacecraft in August 2001 toward the Sun with the specific aim of collecting solar wind particles until September 2004 in ultra pure wafers of silicon and other materials. Despite an unscheduled crash to the Earth, the integrity of some of the samples was preserved.

The color indicates the temperature – the bluer the hotter, the redder the cooler. The extent of red shift tells how fast a star is traveling away from us, the blue shift, toward us.

> The tomb of Christian Doppler (1803–1853) is to be found at the entrance of the Venetian island cemetery of San Michele. He died, aged 49, of pulmonary disease in Venice, then part of the Austrian empire.

The existence and position of emission and/or absorption Fraunhofer lines defines the chemical composition of the star or that of other bodies or clouds in the way. Table 4.8 gives examples of the chemical composition of some stars.

For those stars that are double, the changes in appearance and shift of the spectroscopic lines allow calculation of many parameters – the speed of rotation of the components around each other, the direction of rotation, the size of each, the angle of the plane to us, and so on, much of which is deducible using fairly simple known physical laws.

Similarly, the spectral shifts are useful in determining the properties of galaxies. Observations are not just color as we see it, since shifts apply to all parts of the electromagnetic spectrum such as radio waves and have brought us information about the angle of the plane of the galaxy and the speed of rotation of component stars at various positions. Amazingly, such information indicates the existence of some other material or force that supports the suggestion of "dark matter" that might help to solve the

Table 4.8. The chemical composition of spectral classes.

Spectral class	Temperature, K	Color	Color Index	Dominant lines	Examples
O	50,000–28,000	Blue	–0.3	He II	εOri
B	28,000–10,000	Blue-white	–0.2	He I	Rigel;Spica
A	10,000–7,500	White	0.0	H	Vega;Sirius
F	7,500–6,000	White-yellow	0.3	Metals; H	Procyon
G	6,000–4,900	Yellow	0.7	Metals; Ca II;	Sun;αCen A
K	4,900–3,500	Orange	1.2	Ca II; Ca I	Arcturus
M	3,500–2,000	Red	1.4	TiO; Ca I	Betelgeuse

mystery of why we can only account for a small proportion of the known universe from baryonic observations.

Baryonic matter is the "solid stuff," visible or detectable by the eye, telescopes, or any other conventional means. It is composed of quarks, essentially, the content of the nucleons protons and neutrons. The composition of the vast majority of the rest of the universe is unknown but hypothetical WIMPS (weakly interacting massive particles) and neutrinos as possible candidates for WIMPS have been put forward. Axions are hypothetical, elementary particles of very small mass that have been proposed as candidates for the composition of dark matter.

Information on the ancestry of our own galaxy, The Milky Way, has been gleaned from ESO's Very Large Telescope (VLT) in Chile. A large spectroscopic survey of some of our surrounding dwarf galaxies has measured the amount of iron in over 2,000 individual giant stars in the Fornax, Sculptor, Sextans, and Carina galaxies. Spectra were obtained using FLAMES, the Fibre Large Array Multi-Element Spectrograph, mounted on Kueyen, the Second Unit Telescope of the VLT. Light spectra of up to 130 objects within an area almost the size of the full Moon can be accumulated in a single observation.

The results of spectroscopy provide much information to feed the predictions or imagination of those pursuing astrobiology or exobiology. Does life exist elsewhere in our universe and how could it have started here in the first instance? As stated previously, the early stages of the nuclear fusion of helium creates beryllium-8, a very unstable form of the element. Being unstable it breaks down and decays into helium almost instantaneously. But sometimes the beryllium-8 bumps into another helium atom with just the right amount of energy to form carbon. Once carbon is formed then we can exist! And how could we know all this without spectroscopy?

The half-life, the time it takes to reduce the amount to half of its original value, of beryllium-8 is a staggeringly small 5×10^{-14}s!

CHAPTER FIVE

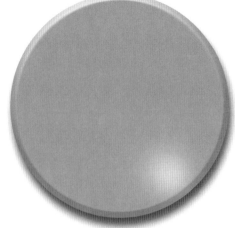

Spectral Techniques

T. Buick, *The Rainbow Sky*, Patrick Moore's Practical Astronomy Series,
DOI 10.1007/978-1-4419-1053-0_5, © Springer Science + Business Media, LLC 2010

Capturing Spectra

Obtaining and recording your own spectra may seem out of the reach of mortal persons, but not so! Maybe capture of the pin-sharp, high-resolution, well-separated multitude of Fraunhofer lines might be a little ambitious to begin with, but satisfying views and photographs can be obtained that will indicate that the colors and lines are actually there. The theories we've been discussing, then, become a great deal more believable. Sources of light are different, and the light itself from the many sources is really different, often dramatically so!

Ever since Newton in the seventeenth century described the separation of the constituents of light by a prism, observers have been fascinated by such brilliant and vivid colors. The red, orange, yellow, green, blue, indigo, and violet hues can be seen everywhere – water droplets, rainbows, jars full of water, broken glass. The photograph (*see* Fig. 5.1), taken many decades ago with a very old film camera, has captured the "rainbow" from a dew drop on the fence.

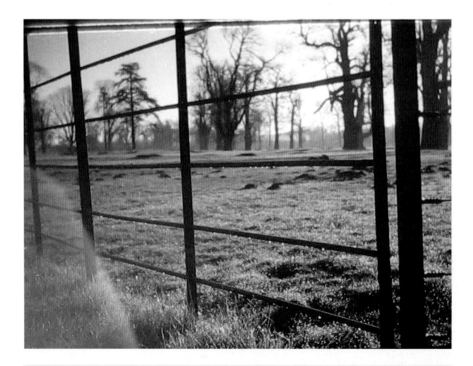

Fig. 5.1. A rainbow from a winter's dew drop.

Prisms

A prism is an obvious piece of equipment to start spectral viewing at home. Prisms are handy and can be salvaged from various redundant optical instruments or purchased quite cheaply. Place a prism at an angle such that its faces/edges are perpendicular to the direction of the Sun's light. Hold in place if necessary using clay or tape. Hold or tape a white card to the side and rotate the prism until the best spectrum is seen. DO NOT LOOK DIRECTLY AT THE SUN OR ALLOW THE LIGHT FROM THE PRISM INTO YOUR EYES. SERIOUS DAMAGE COULD RESULT!! Take a photograph using a compact digital camera (almost any camera will do) on a stand or other support and setting it to "close-up" or "macro" and delayed shutter release to avoid camera wobble. The picture will be simple but attractive, especially to children (*see* Fig. 5.2). However, the absorption lines will likely not be discerned here. With more attention to detail, and patience, clear and well-resolved spectra can be captured, even from stars that exhibit the Fraunhofer lines that define the chemical composition of the incident light. Figures 5.3 and 5.4 demonstrate the variety of star spectra obtainable through the simple attachment of a prism to a camera.

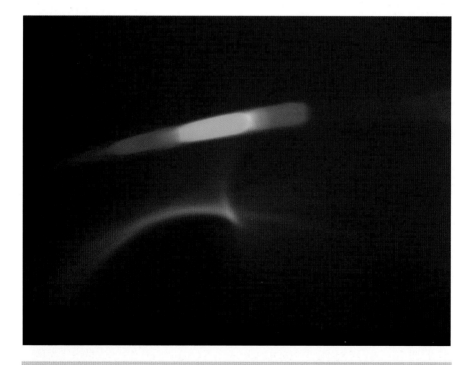

Fig. 5.2. Well defined colors from sunlight through a prism.

Sir Isaac Newton stated that the number of colors in the spectrum was seven, as a result of the common search for a link between music and color. There are seven notes in the octave musical scale, so it was thought that there should be seven colors in the spectrum. Seven was the key principle in alchemy, religious philosophy, and ancient cosmology and was involved in a deeper search for harmony and the meaning of life.

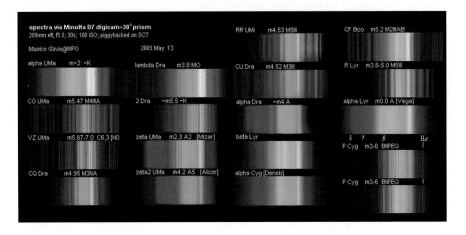

Fig. 5.3. Clear Fraunhofer lines are visible in these camera shots through a prism (Credit: Maurice Gavin).

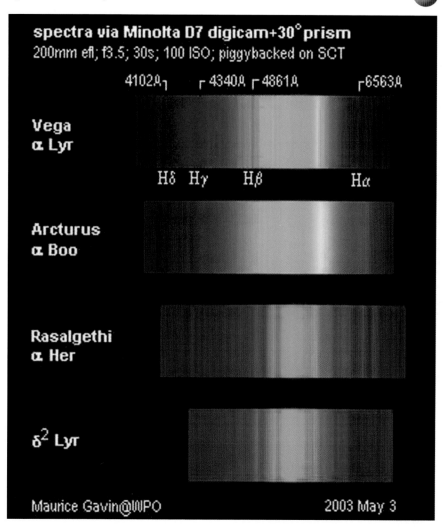

Fig. 5.4. Clear Fraunhofer lines are visible in these camera shots through a prism (Credit: Maurice Gavin).

CDs and Diffraction

For achievable results with little expense, nothing can surpass the use of a compact disc, a CD, which is globally available and cheap. There can be few people who have not by accident left a CD around in sunlight and noticed the stunning array of colors projected onto the wall. Kids love it (*see* Fig. 5.5). The effect occurs because of the coincidental similarity of its surface structure to a diffraction grating. So what is a diffraction grating?

Scientists in the early nineteenth century had to come to terms with the fact that light had the properties of a wave. An experiment carried out by an English scientist Thomas Young in 1800 showed that by passing a beam of light through two closely adjacent slits the results of wave interference on the emerging light could be seen from the ladder-like steps of dark and light projected onto a screen. Where waves overlapped the light was reinforced (constructive interference). Where they were opposite they were destroyed (destructive interference). *See* Fig. 5.6a and b below.

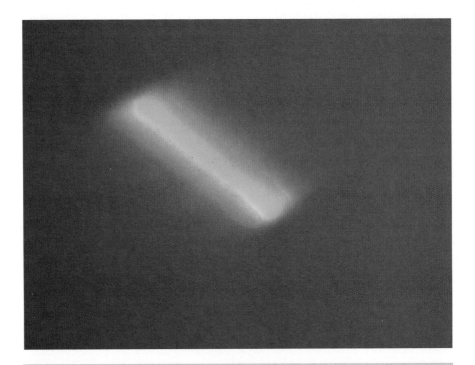

Fig. 5.5. Sunlight colors projected onto a wall from a compact disc.

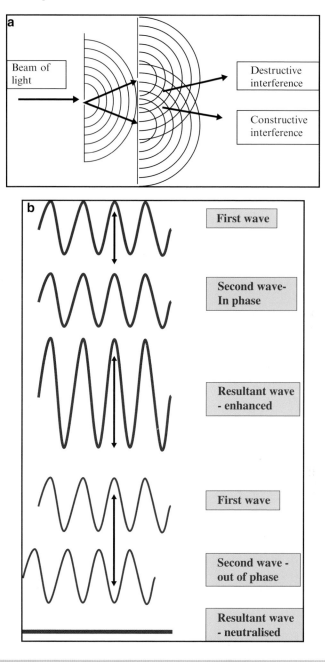

Fig. 5.6. (a) Diffraction of light from two adjacent slits. **(b)** Constructive and destructive interference.

Diffraction was then discovered, or at least brought to the fore, when James Gregory (1638–1675), a Scottish astronomer and mathematician, first observed the phenomenon by passing sunlight through a bird feather a year after Newton (1643–1727) had done the same using a prism. The properties of waves are commonly seen in puddles during rain (*see* Fig. 5.7).

Young's experiment involved just two adjacent slits. Two adjacent, carefully formed grooves or scratches on a transparent surface will serve the same purpose – to allow passage of light waves that interfere. A modern diffraction grating comprises many hundreds of such lines per millimeter to produce a strong spectrum. A CD similarly

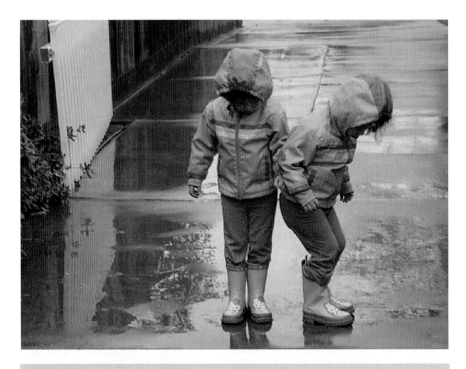

Fig. 5.7. Puddle waves are a good example of constructive and destructive interference. Look in the Appendix of this book for references if you wish to pursue this subject further.

has a huge number of spiral lines inscribed with precision to accept data, only this time on a reflective surface that will cause light wave interference or diffraction by reflection. A beautiful spectrum of vivid colors can be photographed pointing the beam directly into the camera (*see* Fig. 5.8).

It is worth relating a little more history about Thomas Young, who was a "truly original genius," according to Sir John Herschel. At the age of 14 he was fluent, or acquainted with, 12 languages. He became a physician (including medicine and midwifery), was a professor of natural philosophy, became foreign secretary at the Royal Institution, served on a committee to consider the dangers of using gas in London, and became secretary to the Board of Longitude. His work was admired by eminent scientists such as Albert Einstein and Lord Rayleigh. He made important discoveries concerning the structure and mechanics of the eye, the characterization of elasticity (Young's Modulus), capillary action, and surface tension, and functions of the heart and arteries. He also compared the

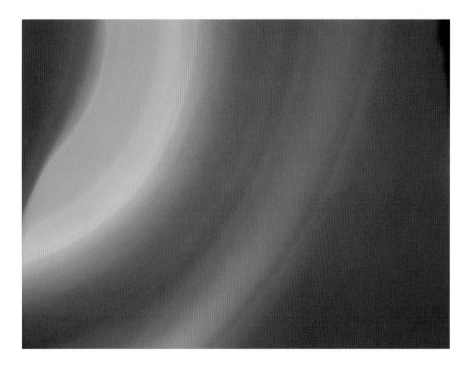

Fig. 5.8. Camera shot of a spectrum of sunlight from a CD.

grammar and vocabulary of 400 languages and contributed greatly to translation of hieroglyphics and the text of the Rosetta Stone – all this, and more, including his contribution to the understanding of light waves and diffraction. He is buried under the chancel of St Giles' church in Farnborough, Kent, England, where the commemorative plaque acknowledges his contributions to humankind (*see* Fig. 5.9).

Jean-Francois Champollion is credited with the final translation of the text of the Rosetta Stone, which was created in 196 BC. The Rosetta Stone contained the same text in hieroglyphics script as was contained in Demotic and classical Greek (a decree from Ptolemy V concerning taxes and the erection of statues) and so became a key to understanding hieroglyphics. The stone has been continuously exhibited in the British Museum in London, England, since 1802, apart from 2 years below ground in the Postal Tube Railway during World War 1.

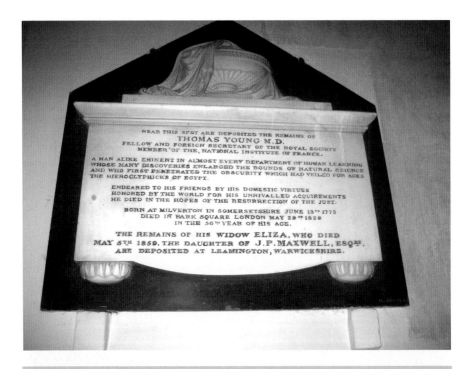

Fig. 5.9. A plaque to commemorate this most eminent of scientists is prominently displayed in St Giles' Church, Farnborough, Kent, England. The full text of the plaque is given in the Appendix of this book.

To stand a chance of seeing the spectral lines within the spectrum one has to be a little more careful. Hold a CD at a glancing angle between you and a light source. DANGER!!! DO NOT LOOK DIRECTLY AT THE SUN OR REFLECT LIGHT DIRECTLY FROM THE SUN INTO YOUR EYES! SERIOUS DAMAGE TO YOUR EYES MAY RESULT. Vary the distance between your eye and the CD. Often, almost touching the CD is best. Not only will the spectrum be seen but, with care, many Fraunhofer lines also. As with the first sight of Saturn and its rings through a telescope, the first sight of a home-produced, lined spectrum is very exciting (*see* Fig. 5.10). This image was taken by holding a digital camera against the edge of a CD. The photograph just captures some of the lines, many more of which could be seen with the eye, although it is not wise to try that. It can be addictive taking pictures of the CD spectra of every source of light available.

The spectrum in Fig. 5.11 was captured using the same simple technique from a sodium street lamp. Figure. 5.12 shows the spectrum captured from the currently recommended low energy domestic light bulb.

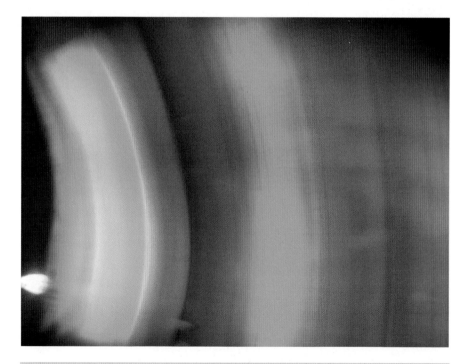

Fig. 5.10. Fraunhofer lines visible from reflected, diffracted sunlight.

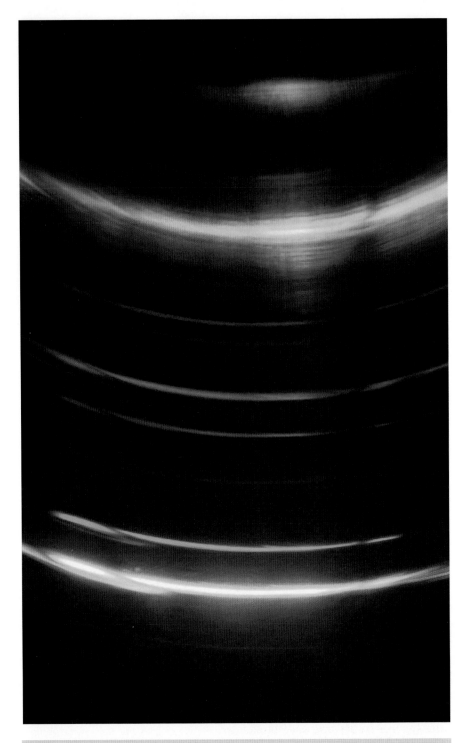

Fig. 5.11. Diffraction spectrum of light from sodium street lamp. Spectral colors characteristic of that particular light source are easily identified.

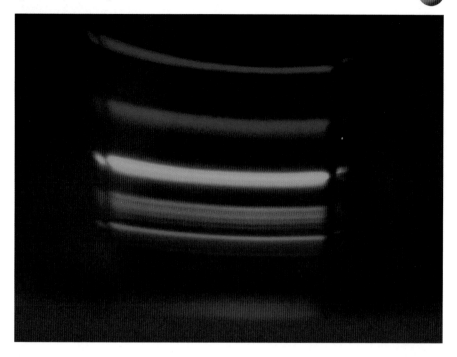

Fig. 5.12. Diffracted light from a low energy domestic bulb.

A cardboard box with two razor blades close together to form a slit, and a sawn off CD, allowed somewhat better pictures. Among some stunning images available using the "cereal box" technique are those produced by Joachim Koppen and reproduced here with his permission. Figures 5.13 and 5.14 are his photographs of the spectrum of the Sun. Joachim writes: "The spectroscope (cereal box) was directed towards the Sun; thus brilliant spectra were visible in the (single lens reflex) camera. A 100 mm telephoto lens was used to give some enlargement of the spectrum. The aperture was fully open – it served no function here. The exposure time on ordinary 200 ASA color negative film is 1/30th of a second."

Longer exposure times are required for objects less brilliant than the Sun. Figure 5.15 is a spectrum of a mercury-vapor street lamp. Figure 5.16 is a composite picture of the second order spectrum. "The yellow doublet (at 5769.6 and 5789.7 A) is clearly resolved." Figure 5.17 reveals the fine lines from mercury and sodium lamps.

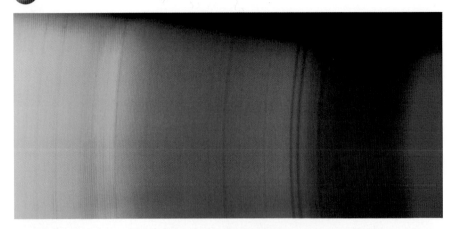

Fig. 5.13. "Cereal box" spectra of the Sun. (Credit:Joachim Koppen) The photographs are enlarged to enable appreciation of the resolution achievable through this simple technique.

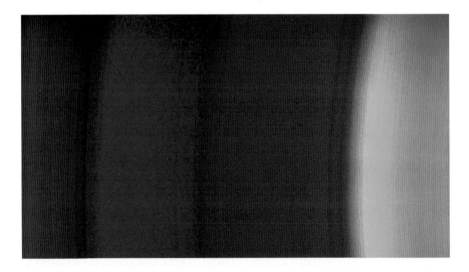

Fig. 5.14. "Cereal box" spectra of the Sun. (Credit:Joachim Koppen) The photographs are enlarged to enable appreciation of the resolution achievable through this simple technique.

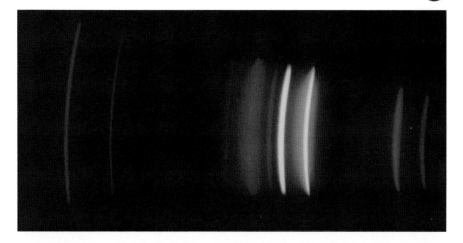

Fig. 5.15. First order spectrum of a mercury-vapor street lamp (Courtesy of Joachim Koppen.)

Clearly, "cereal box" spectroscopes are unique to every "cereal box spectroscopist," as the design depends on the particular household packaging and materials to hand. Figure 5.18 presents the spectrum of a fluorescent light, and Fig. 5.19 a metal halide lamp. (References to Joachim Koppen's and Jerry Xiaojin Zhu's websites are given in the Appendix of this book for further details of these spectra and how to produce them.)

Certainly a CD behaves as a diffraction grating, but how about the real thing? As application requirements become ever more demanding, science and technology, with ever increasing precision and inventiveness, meets those demands and reveals yet more possibilities. A good example is the growing selection of types of grating available. It is surprising how any science topic is likely to become more complex the more it is investigated. With each extra fathom of depth into a subject the complexity usually increases unimaginably. So it is with gratings; the grating equation, glass, film, holographic, ruled, transmission, reflection, orders, substrate, coatings (aluminum, gold, silver, chromium), protective overcoat, groove profile (blazed, sinusoidal) to list but a few buzz words. Fortunately, there

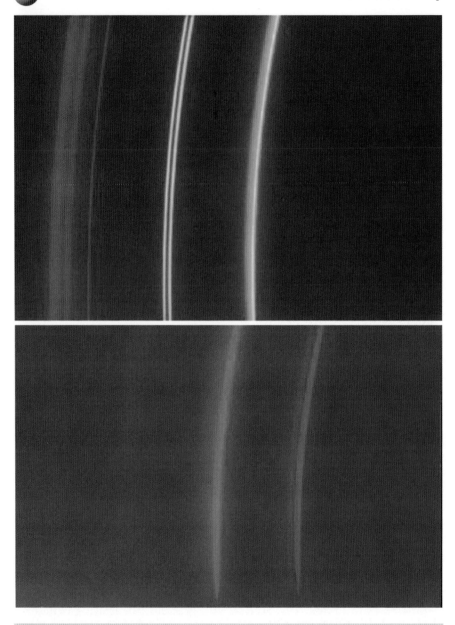

Fig. 5.16. Second order spectrum of a mercury-vapor street lamp (Courtesy of Joachim Koppen.)

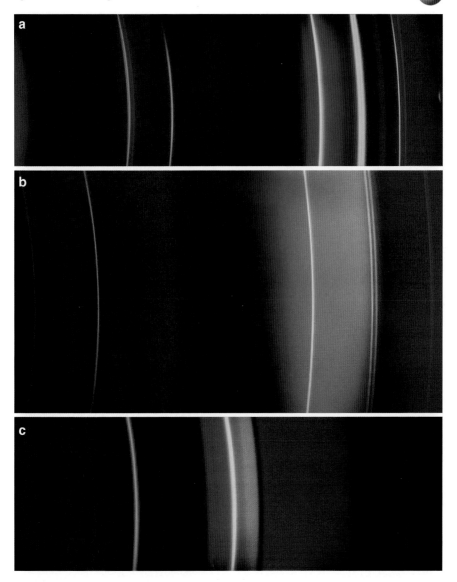

Fig. 5.17. Sharp diffraction lines from mercury and sodium lamps (Courtesy of Joachim Koppen.)

Fig. 5.18. A cereal box spectrum of a fluorescent lamp (Credit: Jerry Xiaojin Zhu).

are some off-the-shelf products that are very easy to use and, although how much you spend can rival the cost of a large telescope, really useful ones can be purchased for the price of a good telescope eyepiece (or much cheaper thin-film products).

The spectrum in Fig. 5.20 was captured using a holographically produced glass grating, blazed (groove shaped) to concentrate brightness in the first order spectrum and with an added cylindrical lens to spread

Fig. 5.19. A cereal box spectrum of a metal halide lamp (Credit: Jerry Xiaojin Zhu).

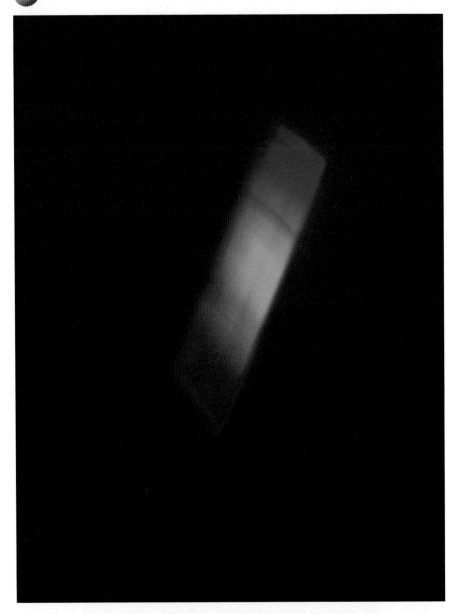

Fig. 5.20. Diffraction grating spectrum of Vega.

the image. The grating was simply screwed into the base of a 25 mm eyepiece, with the cylindrical lens attached to the top, and inserted into a Celestron C8, equatorially driven telescope. The spectrum captured clearly shows the major hydrogen absorption lines in spite of less than ideal seeing conditions.

As every child knows (*see* Fig. 5.21), bright bands of color are often seen when light reflects from a soap bubble or a thin layer of oil floating on water. These are the result of interference. Light waves are reflected from opposite surfaces of such thin films (hence one beam travels further than the other), and constructive interference between the two reflected waves occurs in different places for different wavelengths. So the reflected, diffracted light appears colored. Still frames from movies of light from low energy and sodium lamps as reflected from soap bubbles are displayed in Fig. 5.22.

Spectroscopy is not just a pretty way of collecting alternative pictures of stars, although it is great fun. Amateurs and professionals use them to unfold the secrets of the structure of the universe and its contents. There are many vital applications that depend heavily on the accumulation and analysis of spectra. Few could be more important than the constant vigil for spotting harmful objects flying towards us from outer space.

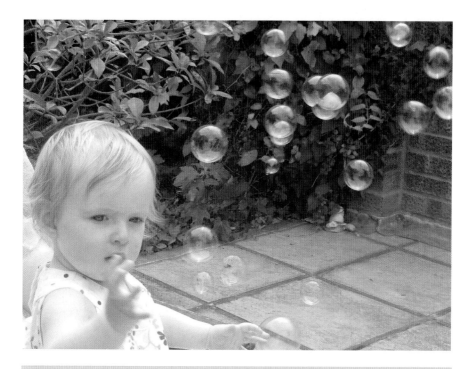

Fig. 5.21. Kids know all about the bubble's colors!

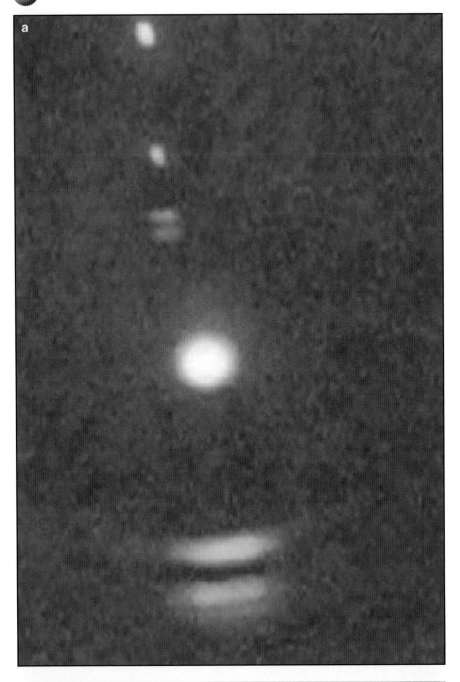

Fig. 5.22. Excellent examples of diffraction for spectators to witness and discuss.

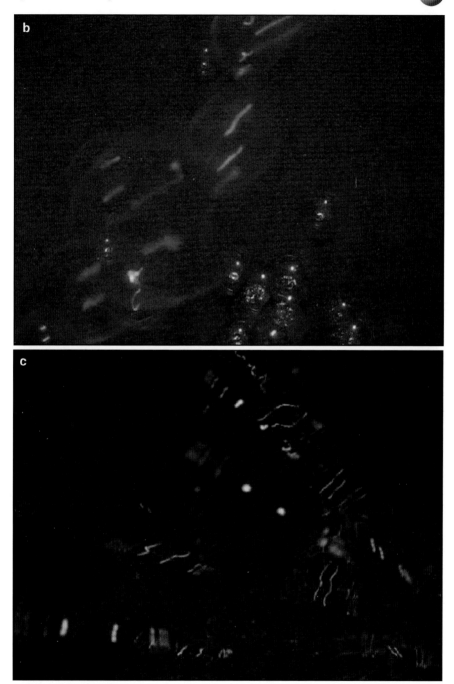

Fig. 5.22. (continued)

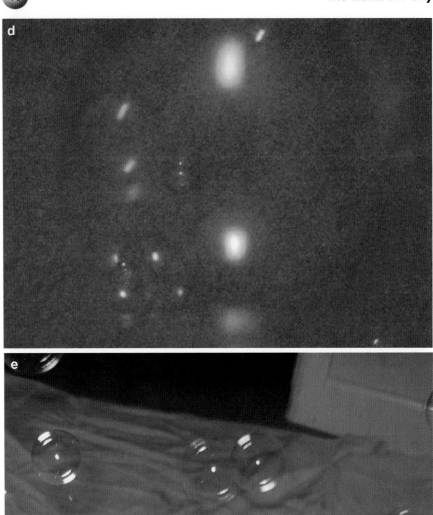

Fig. 5.22. (continued)

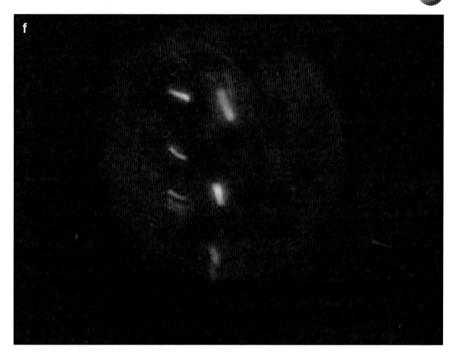

Fig. 5.22. (continued)

Only through careful study of the structure of a lump of rock, when far enough away, might disaster be avoided. By comparison of its spectrum of reflected light with that of objects of known composition the structure might be deduced. Knowing whether it is a solid lump of iron or a loose aggregation of smaller pieces of rock could contribute greatly to the design of effective countermeasures. Another ambitious project, Marco Polo, is now on the drawing board at UK EADS Astrium and OHB Germany with the objective of collecting material from an asteroid and returning it to Earth. The more knowledge gained about the composition of asteroids the more we can research into possible methods of deflection.

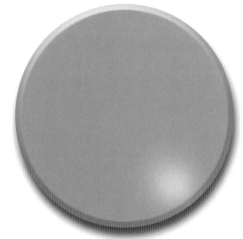

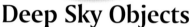

CHAPTER SIX

Deep Sky Objects

A wealth of information has been accumulated about stars, gleaned to a great extent from their characteristic spectra, and it is very satisfying to know the class and composition of those sparkling bright spots we see in the night sky. But there is much more to investigate up there – gas clouds, explosions, spirals, weird shapes. Even though the objects are often thousands or millions of light years away the apparent diameter of many of these is greater than that of the Moon as we see it from Earth. And they all have fantastic colors (see Fig. 6.1). But can we see those colors; and what do they mean?

It is an exciting adventure to explore those wonderful exotic smudges in the sky. Look at the dramatic colors shouting out at you from some of the breathtaking pictures available of deep sky objects, nebulae, and galaxies, and you will hear yourself say, "You cannot be serious!" Although a full discussion of how color photographs are captured, and how close they are to what the eye perceives, would require another book, a few words here might be helpful.

Attempting to photograph a distant object in high-resolution full color in one fell swoop is not the best way to go. At the very least it is extremely inefficient time wise, and accumulation of detail could well be at the expense of overexposing the brighter features. The photographic process

T. Buick, *The Rainbow Sky*, Patrick Moore's Practical Astronomy Series, DOI 10.1007/978-1-4419-1053-0_6, © Springer Science + Business Media, LLC 2010

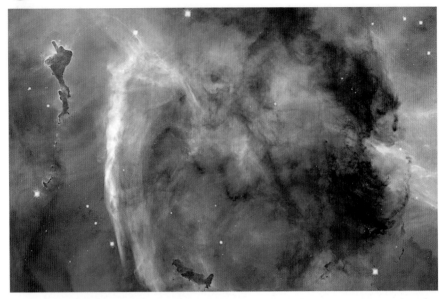

Fig. 6.1. Light and shadow in the Carina or Keyhole Nebula (NGC 3372), about 8,000 light years away and near the explosive variable star Eta Carina (courtesy of NASA/STSci/Hubble space telescope).

begins with monochrome images captured through each of three special filters – red, green, and blue. Monochrome detectors are more sensitive than colored ones. The intensity of each of the colors in the picture is then known. Combining the three color pictures creates the full-color image that is adjusted for overall intensity by comparison with an unfiltered monochrome photograph.

As mentioned in other chapters, some images are reproduced to be as close as possible to natural color, the colors we can see, and some processed to enable us to see otherwise invisible colors, such as infrared and ultraviolet; some are done deliberately using contrasted colors to present variations in physical, chemical, or structural properties. So, we have true or natural color and false color.

Abell (1689) (see Fig. 6.2) was captured, in part, by the Spitzer space telescope, and so it is, by its very nature, a false color image from infrared, recreated in glorious golden and amber hues. Each of the huge number of smudges hosts billions of solid and gaseous structures, many being millions of light years across. The huge gravity created by the trillions of

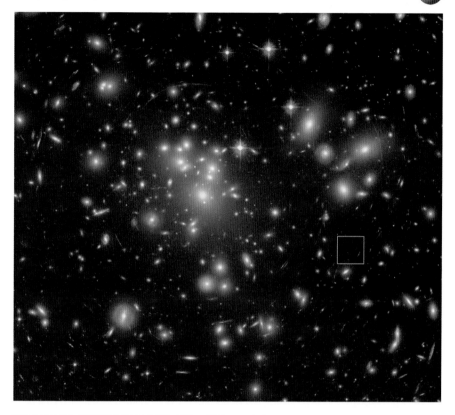

Fig. 6.2. Distant galaxy cluster Abell 1689 (credit: NASA/ESA/L. Bradley and H. Ford, Johns Hopkins University; R. Bouwens and G. Illingworth, University of California, Santa Cruz).

stars acts as a lens that bends and magnifies light from the galaxies behind that appear as arc-shaped objects around the cluster. It was Fritz Zwicky who first suggested in 1937 that the Einstein Effect, a "warped space" prediction of Einstein's general theory of relativity, could allow galaxy clusters to act as gravitational lenses. This was eventually confirmed by observation of the "Twin Quasar" Q0957 + 561 in 1979.

> Edwin Hubble, in the 1920s, was the first to observe that the so called Extra Galactic Nebulae were in fact composed of a myriad stars. Since then the name was reduced to "galaxy."

As a complement to Spitzer, Fig. 6.3 shows a Hubble space telescope deep field false color image of an area about one fortieth the size of the Moon.

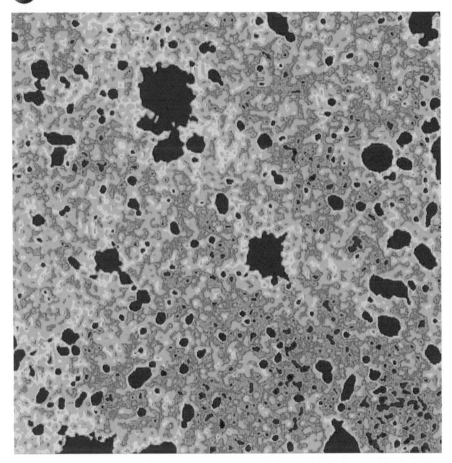

Fig. 6.3. Hubble deep field false color image showing distant galaxies and stars in red (credit: NASA/Dr. Michael S. Vogeley, Princeton University Observatory, now affiliated with Drexel University).

The red regions are galaxies and stars, the smallest red patches being four billion times fainter than the human eye can see! Within our own galaxy, the Milky Way, a multitude of masses of stars, gasses, or a mixture of the two have been discovered and photographed by land-based and space-based telescopes, and presented in true and false colors. But what of other galaxies?

It is easy to be overwhelmed by the sheer number of smudges in the night sky, and the stronger and more capable the telescope the more awesome the vision becomes. Someone had to make a start to bring order to the nomenclature. Charles Messier (1730–1817) was a French astronomer and comet hunter appointed by the Paris Observatory to search for the return of Halley's Comet in 1757. To avoid confusion with other apparent space clouds he compiled a list of nebulous objects, the first being M1, now known as the Crab Nebula. Many of the (now) 110 objects he discovered himself. A later full index of star clusters, nebulae, and galaxies, the New General Catalog, was compiled in 1888 by the Danish astronomer John Dreyer and now includes thousands of entries that virtually cover the whole sky, although modern instruments continue to discover new entities. The Crab Nebula, Fig. 6.4, is therefore designated

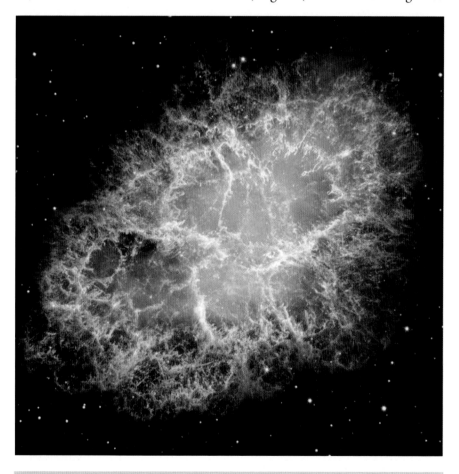

Fig. 6.4. A composite image of the Crab Nebula from the Chandra X-ray telescope, Hubble space telescope, and the Spitzer space telescope (credit: NASA/ESA/CXC/JPL-Caltech, J. Hester and A. Loll of Arizona State University, R. Gehrz of the University of Minnesota, and STSci).

M1 and NGC 1952. An update, the Index Catalog, was published in two parts in 1895 and 1908, and the identity IC is used. Sir Patrick Alfred Caldwell Moore produced a list, the Caldwell Catalog, of 109 deep sky objects (not in Messier's list) of help to amateur astronomers possessing a small telescope.

Deep sky discoveries are being made all the time. One of the most ambitious of international astronomical projects is the Sloan Digital Sky Survey (SDSS), a detailed, three-dimensional optical survey of more than a quarter of the sky. (See the Appendix in this book for more information.) The 2.5-m telescope situated on Apache Point, New Mexico, hosts a 120 megapixel camera and a pair of spectrographs. The first phase of the SDSS, completed in June 2005, detected nearly 200 million celestial objects and measured the spectra of more than 675,000 galaxies, 90,000 quasars, and 185,000 stars. The second phase is ongoing, and a third is expected to follow. From these we hope to learn a great deal about the nature of the universe.

Galaxies

Galaxies come in a range of sizes. Their masses of a few million to many trillions of stars, combined with gas and dust, are mostly a few thousand to a few hundred thousand light years across. It is not surprising that these swirling masses form themselves into different shapes, as observed by William Herschel and his son John from 1781 to 1847. William Parsons, the third Earl of Rosse, described the spiral nature of some galaxies in 1845, and Edwin Hubble in 1925 designed the first classification of the types of galaxies, which was known as the Tuning Fork diagram due to its shape (see Fig. 6.5).

Elliptical galaxies: Smooth light distributions denoted by the letter E followed by an integer *n* to denote the degree of ellipticity observed.

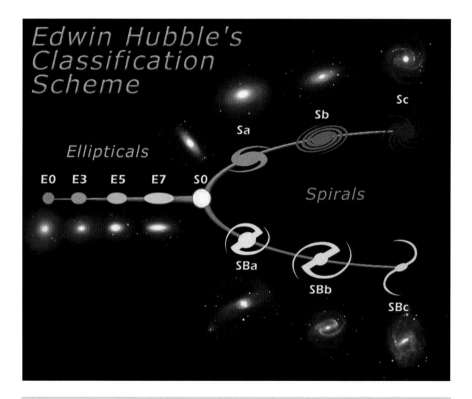

Fig. 6.5. Edwin Hubble's galaxy classification scheme (credit: NASA). Three major morphological classes are defined: elliptical, spiral, and lenticular.

Spiral galaxies: Flattened disks with stars that form a spiral structure and a central concentration of stars that form a bulge. These are denoted by the symbol S. About half of all spirals also appear to have a bar-like structure stretching out from the bulge. Barred spirals are denoted by the code SB.

Lenticular galaxies: Bright central bulges surrounded by disks but without any visible spiral structure. Designated as S0, Lenticulars are not very active star-forming entities.

These classifications can be extended to include other irregular types also.

Gérard de Vaucouleurs extended the Hubble classification in 1959, in particular to the spiral morphology, as rings and lenses were important components. He therefore introduced a more detailed description of the three components of spirals.

Bars: The designation SA was added, to complement Hubble's SB, to denote the absence of bars, and SAB for weakly barred spirals. Lenticular galaxies were also classified as barred SB0 or unbarred (SA0) reserving S0 for where it was not possible to tell whether there was a bar or not.

Rings: Galaxies with ring-like structures (r) and without (s), and with (rs) for uncertainty.

Spiral arms: As in Hubble's original diagram, spiral arms are described according to the tightness of the arms but extended to include additional classes. Sd (SBd) referred to diffuse broken arms and a faint central bulge; Sm (SBm) galaxies are irregular in appearance but have no apparent central bulge; Im refers to a highly irregular galaxy.

As an example, a weakly barred spiral galaxy with loosely wound arms and a ring is denoted SAB(r)c.

More subcategories are described using numerical values in both the Hubble and de Vaucouleurs systems to designate the degree of conformity to the standard shape, and a reference is given in the Appendix of this book for further designation definitions and the Yerkes or Morgan scheme for classification using the spectrum of component stars.

The Andromeda Galaxy, M31, and the Large and Small Magellanic Clouds are the only galaxies visible to the naked eye and have probably been known since prehistoric times. Certainly M31 was noted by medieval Persian astronomers around A.D. 900 and fully described and cataloged by the Persian astronomer Al Sufi in 964 A.D. It was only following the invention of the telescope that many more galaxies came to light. The next three to be discovered were:-

The Triangulum Galaxy M33 – seen by G.B. Hodierna, an Italian priest – astronomer before 1654.

M32 – a companion to Andromeda, by French astronomer Legentil in 1749; and

M83 – the first galaxy to be discovered beyond the Local Group, by Abbé Lacaille in 1752.

Following Messier's search for anything that could be mistaken for a comet, came a number of discoveries, the first find being M49, a giant elliptical galaxy in the Virgo Cluster observed in 1771.

How many are there of each; and are there more? A project directed by Dr Chris Lintott of Oxford University in England is asking members of the public to log onto a website to view a million-strong database of galaxies to do what computers are unlikely to do as well as humans – to decide what type each galaxy is. The project is called Galaxy Zoo (see the Appendix of this book for more information). The response has been great, and much has been gained from the exercise, such as Hanny's Voorwerp Nebula being discovered. Further inspection of the objects can now be asked of the lay public, and astronomers will assess more detailed information, such as the number and description of spiral arms. Who knows what exciting discoveries might be made in addition to its prime objectives?

Galaxies can exist alone but are mostly found in groups. Our own galaxy, the Milky Way (see Fig. 6.6), and the Andromeda Galaxy are the dominant ones in a cluster of around 20 in our "Local Group," although

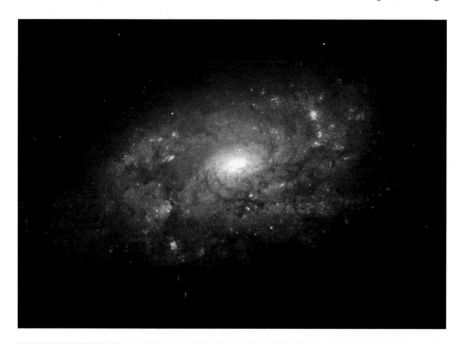

Fig. 6.6. NGC 3949, a spiral galaxy similar to our own Milky Way (credit: NASA/ESA/Hubble Heritage/STSci/AURA).

some are very small. Eight million light years away is another cluster, the Sculptor group, and the Messier 81 group lies 12 million light years away. The nearest large cluster is Virgo at a distance of 65 million light years, and the furthest away that we know of (a whopping nine billion light years away) presents us with a view of the very young universe. But these are not benign, well-ordered agglomerations of rocks, gasses, and fiery entities.

If we could stand back and view the cosmic movie speeded up with super stereo sound as well, what a vision that would be! What violence! Galaxies would be blasting their way into each other, shock waves would be crashing into vast gas clouds to trigger star formation, and outer cold and icy rocky bodies of distant solar systems would be flung chaotically out of orbit to smash into any unfortunate planet that would anyway be roasted as its star succumbed to the imbalance of forces at the end of its life. Even our own Milky Way and Andromeda are heading for a collision in a few billion years time, a presage of which is illustrated by the Mice (NGC 4676) colliding galaxies (see Fig. 6.7).

Stunning images of galaxies have been captured and presented in true or false colors to indicate what they look like to the human eye or to bring out some highlights or properties. Here are a few examples.

M100 is a spiral galaxy. The camera that captured this image has the ability to resolve individual stars in other galaxies and hence allow spectroscopic analysis of their composition (Fig. 6.8).

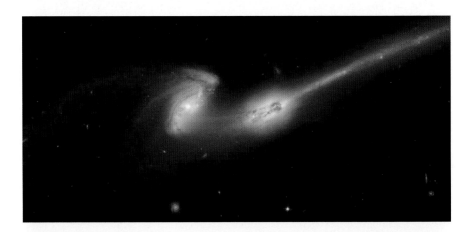

Fig. 6.7. The Mice colliding galaxies 300 million light years away in the constellation of Coma Berenices (credit: NASA, H. Ford [JHU], G. Illingworth [UCSC/LO], M. Clampin and G. Hartig [STSci], the ACS Science Team, and ESA).

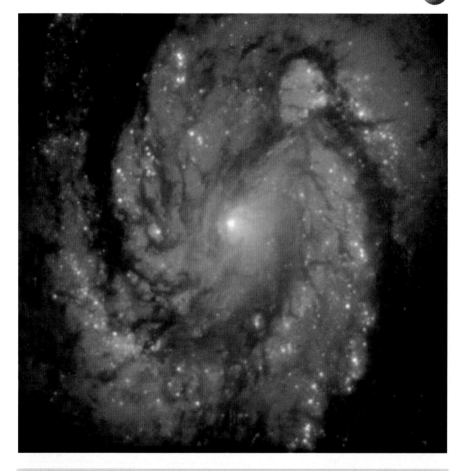

Fig. 6.8. M100, NGC 4321 (credit: NASA/STSci).

Location: A member of the Virgo cluster in Coma Berenices
Distance: 60 million light years
Telescope: Hubble space telescope
Camera: Wide field and planetary camera
Description: Spiral galaxy tilted nearly face-on and among the first spirals
    to be discovered
Image: Taken on December 31, 1993, through red, green, and blue filters
    to create a true color picture. Blue corresponds to the light from young
    and massive stars recently formed along the spiral arms.

The Antennae galaxies are a great example of what will happen when our own galaxy meets Andromeda, at present with a closing speed of thousands of km per second. Current calculations are not sufficiently precise to say whether it will initially be a head-on collision or a glancing blow. Either possibility is likely to cause catastrophic consequences, even though most of the bulk of galaxies is space. Representing our stars by the size of a pinhead, each star, on average, would be many km away from its nearest neighbor. No need to panic about that eventuality, however; it is better to worry more about the end of Earth for humankind as we know it, which will occur as our own star enters the last phases of its existence. The two Antennae spiral galaxies that smashed together a few hundred million years ago caused a spectacular starburst activity that led to the formation of billions of stars mostly in clusters, many of which will not last more than ten million years (Fig. 6.9).

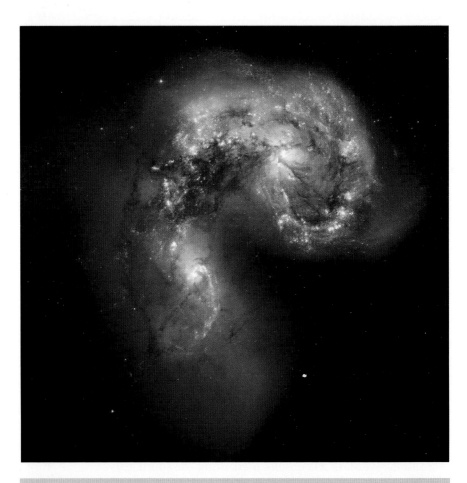

Fig. 6.9. Antennae galaxies, NGC 4038/4039 (credit: NASA/ ESA/Hubble Heritage Team/STSci/AURA-ESA).

Location: The constellation Corvus
Distance: 63 million light years
Description: Merging of two spiral galaxies NGC 4038/4039
Telescope: Hubble space telescope
Camera: Advanced camera for surveys (ACS)
Image: True color

The name "Antennae" comes from the two arms, tidal tails that formed around 200 million years ago, best seen within lower-resolution ground-based telescopes. Brilliant blue regions are stars formed during the crash, some hosting huge numbers and referred to as super clusters that may survive to form regular globular clusters.

The small and large Magellanic Clouds can easily be seen in the southern sky and are so named because they were noted by the Portuguese explorer Magellan. Some of the brightest known star clusters are to be found in the LMC, and the hottest blue stars can be brighter than a million of our Suns. The image is a detailed view of a particular star-forming patch, LH 95, the largest stars of which generate high levels of ultraviolet radiation that heat the surrounding interstellar gas to give a bluish hue of glowing hydrogen. Denser parts of the region present dark strips or reddish hues due to the absorption of blue light from emanations behind it (Fig. 6.10).

Ferdinand Magellan (1480–1521) learned cartography and navigation at an early age and joined expeditions to India and Morocco, where he was badly wounded. Magellan took up Spanish nationality because Charles the First supported his voyage to the Spice Islands with five ships and promised him a share of the spoils, which King Manuel of Portugal would not. Magellan found a passage that went from the Atlantic Ocean through to the Pacific, thence known as the Magellan Straits. In 1521 he discovered the Philippines, where he was killed by natives on the island of Mactan.

Location: The small and large Magellanic Clouds are satellite galaxies of our Milky Way.
Distance: They are "only" around 200,000 light years away. Compare that with the fact that our Milky Way is 100,000 light years across.
Telescope: Hubble space telescope
Camera: Advanced camera for surveys
Image: A composite of visible and infrared filters renders the doubly ionized hydrogen bluish and also helps to distinguish hot bright blue stars from cooler red ones.

NGC 1132 is a distant elliptical galaxy possibly formed as a result of the merger of a group of galaxies. It is often referred to as a "fossil group of

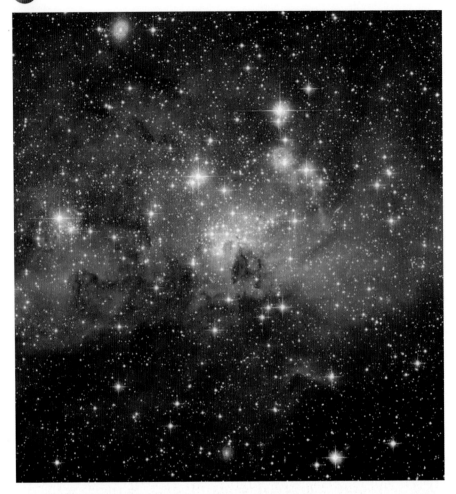

Fig. 6.10. Star-forming region LH 95 in the Large Magellanic Cloud (credit: NASA/ESA/Hubble Heritage Team/STSci/AURA-ESA).

galaxies" because it is thought to contain a huge amount of dark matter. Its strong X-ray signature from hot gas is typical of that found in galaxy groups. The blue/purple in the image is the X-ray glow from hot, diffuse gas (Fig. 6.11).

Location: In the constellation of Eridanus
Distance: About 318 million light years
Telescopes: Hubble space telescope and Chandra X-ray observatory
Camera: Advanced camera for surveys

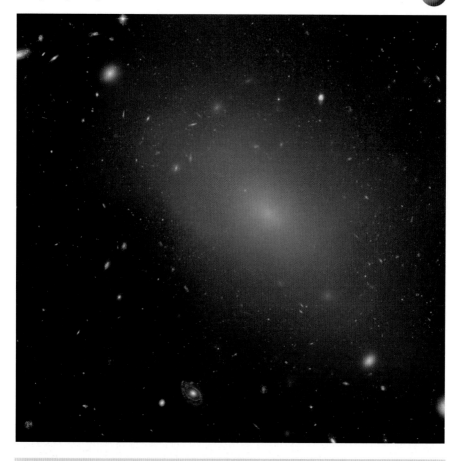

Fig. 6.11. Combined image of NGC 1132 from Chandra X-ray and Hubble space telescope (credit: NASA, ESA, M. West [ESO, Chile], and CXC/Penn State University/G. Garmire, et al.).

Image: A composite through green and near infrared filters from Hubble combined with an image from the X-ray observatory. As if to underline how crowded the cosmos is, many more galaxies, both in the neighborhood and much more distant, can be seen in this image.

Although this dwarf galaxy (see Fig. 6.12) is not a popular astrophotography target for amateurs, it is well studied by professionals, due to its proximity and hosting of two prominent superstar clusters of different ages. Wolf-Rayet stars are also present in the maelstrom. The majority of the clusters seem to have been produced in an energetic starburst that began

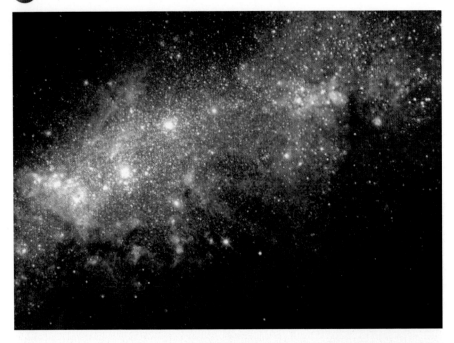

Fig. 6.12. Massive star birth activity in NGC 1569 (credit: ESA/ NASA/P. Anders of the Göttingen University Galaxy Evolution Group, Germany).

around 25 million years ago. Information gleaned can provide clues to star birth and demonstrates that star formation in dwarf galaxies occurs in short bursts, not continually. Its blue-shifted spectrum indicates that it is moving toward Earth, whereas most other galaxies are moving away due to the expansion of the universe.

Wolf-Rayet (WR) stars are massive stars, around 20 times that of the Sun, approaching the end of their lives. The products of the nuclear reaction create a very thick atmosphere comparable to the size of the star itself (which is why we can only "see" the atmosphere, not the star) and generate strong winds that eject huge quantities of material to move the star toward extinction even faster.

Location: In Camelopardalis
Distance: 7.8 million light years
Telescope: Hubble space telescope

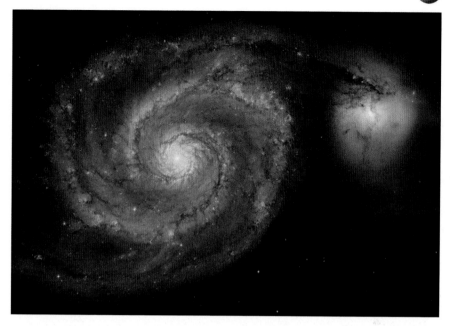

Fig. 6.13. The face-on spiral galaxy M51 (the Whirlpool Galaxy) (credit: NASA/Hubble Heritage Team [STSci/AURA]).

Camera: Wide field and planetary camera 2

Image: A combination of four different exposures through the following filters: a wide ultraviolet (shown in blue), a green filter (shown in green), a wide red filter (shown in red), and a hydrogen alpha filter (also shown in red).

M51 actually comprises two galaxies, labeled as M51A and M51B, or NGC 5194 and NGC 5195. It is studied especially for the structure of its prominent spiral arms, which host sites of the birth of massive and luminous stars, and galaxy interactions between the two companions (Fig. 6.13).

Location: Constellation of Canes Venatici

Distance: 31 million light years

Telescope: Hubble space telescope

Camera: Wide field and planetary camera 2

Image: A composite image showing visible light and emission from glowing hydrogen

Although Charles Messier was the first to see it in October 1773, this galaxy is most famous for being the first galaxy to be recognized as a spiral by Lord Rosse in 1845. The existence and history of Lord Rosse's huge telescope, the Leviathan, is fascinating. The telescope is located within the grounds of Birr Castle, County Offaly, in Ireland, which is home to the seventh Earl of Rosse. The following extract of the Birr Castle website (see the Appendix of this book for reference) is reproduced with permission to describe the history of the telescope (see Fig. 6.14). Figure 6.15 reveals the enormity of the telescope tube (and the difficulty of photographing such contrasting colors).

"During the 1840s and starting from virtually first principles, the third Earl of Rosse designed and had built the mirrors, tube and mountings for a 72-in reflecting telescope, which was the largest in the world at that time and remained so for three quarters of a century. With this instrument, situated near the middle of Ireland, Lord Rosse was able to study and record details of immensely distant stellar objects and to provide

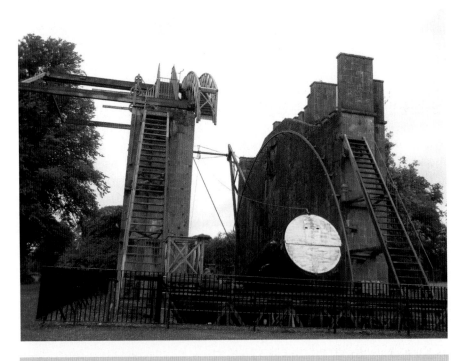

Fig. 6.14. Restoration almost complete, August 2008.

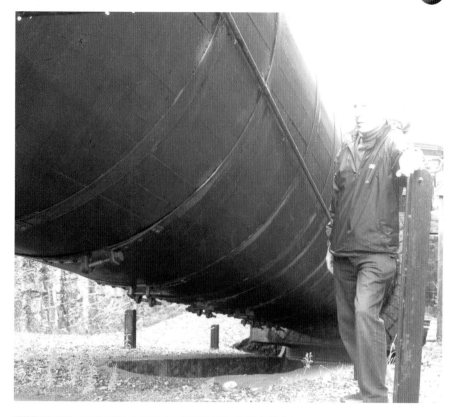

Fig. 6.15. Appreciation of the size of the telescope tube is brought home when one stands next to it.

evidence that many of these mysterious nebulae were actually galaxies located far outside our own.

"The future third Earl of Rosse, William, was given the title of Lord Oxmantown from his birth on the 17th of June, 1800. He was the eldest of three sons, all of whom were educated at home in Birr by private tutors rather than being sent to public schools in England. This type of education may have greatly helped William's natural instinct for engineering, practical aspects of which surrounded him on the Birr Castle demesne. Private tuition gave way to a university education when William was 18 years old. He then went to Oxford College, and graduated with a first class honors Mathematics degree in 1822. William joined the Royal Astronomical Society in 1824 and represented Co. Offaly (at that time known as King's County) from 1823 to 1834 in the House of Lords, supporting both Catholic Emancipation and the Reform Bill. He retired from politics to pursue his scientific and engineering interests, and in 1836 he married

Mary Wilmer-Field, a wealthy heiress from Yorkshire. The financial security from this marriage, as well as the ownership of Birr Castle (which his parents granted him before they left Ireland to live in Brighton, England) allowed him to realize his scientific ambitions and plans. In 1828 he published his first experimental results regarding the grinding and polishing of telescope mirrors in the *Edinburgh Journal of Science*. At the outset of his career William decided to publish all his results, in marked contrast to many other telescope makers. In 1831, he joined the British Royal Society (and was its president from 1848 to 1854).

"The first major telescope built by Lord Rosse was based around a 36 in. (91 cm) mirror. After experimenting with copper and tin alloys for the metal reflecting surface and developing his own steam-driven mirror grinding machine, this telescope was completed in 1839. It was supported in a frame which was a modification (and improvement) of a design by William Herschel. A circular track allowed this telescope to reach most areas of the sky. Despite this telescope being "home-built" (the mirror was cast in the grounds of Birr Castle) two contemporary expert observers confirmed its quality and performance, which allowed significantly increased resolution of stellar objects, one of them (Dr. Thomas Robinson of the Armagh Observatory) calling it the most powerful instrument of its time. Lord Rosse used the 36 in. telescope to study the Moon in greater detail than was previously possible, under a magnification of 900 times. He also studied star clusters and nebulae (fuzzy patches of gas), which earlier telescopes were incapable of resolving to any great degree.

"Lord Rosse spent over 3 years building his next large telescope, "The Leviathan of Parsonstown," based around a 72 in. (183 cm) mirror, with which he hoped to confirm that some of the nebulae contain stars, hints of this being given by the 36 in. telescope. This telescope was basically a 17 m tube, suspended between two 15-m-high walls, with a network of supports which enabled the tube to be moved freely in a vertical direction, but restricted horizontal movement to the extent that the total viewing time on any particular object varied from about 50 min for an object at the equator to about 2 h for an object at the highest level. The vertical movement was contained between the lowest level of 15° elevation and the highest level of 15° beyond the vertical, i.e., a total vertical travel of 90°. Again the mirror was cast in the grounds of the Castle, using three large crucibles and resulting in a mirror that weighed over 3 tons. Construction of the telescope and mountings took over 2 years, but finally, on 15 February 1845, the weather cleared long enough for a short viewing of the double star Castor, which confirmed the potential of the new telescope. In April 1845, Lord Rosse was able to deduce the spiral nature of the M51 nebulae."

After the death of the fourth earl in 1908, the giant telescope fell rapidly into disrepair. However, full restoration has now been achieved and it is well worth a visit.

Nebulae

The space between stars in the universe is not empty. It is filled with interstellar material – gasses and dust. The gas, with temperatures ranging from 10 to 10 million degrees Kelvin, is predominantly hydrogen and helium with small quantities of other gasses such as carbon monoxide. The dust is composed mainly of silicate material with lesser amounts of carbon, titanium, and calcium. A variety of organic molecules have been detected, such as polycyclic aromatic hydrocarbons (they may require final confirmation), benzene-based compounds, amino acids, methanol, and ethanol, the alcohol in your beer! The study of interstellar gas-phase and particle surface chemistry is still relatively new, but more than 120 different molecular species have been detected in the inhomogeneous interstellar medium of dense clouds and sparser regions. Isotopes (deuterium, carbon-13, and oxygen-18), common Earthly gasses (water, ammonia, formaldehyde – as well as the entities referred to above), positive ions (H_3^+, HCO^+, H_3O^+), radicals, and unusual three-membered rings (C_3H, C_3H_2) have been identified. (Ref: Eric Herbst, Chem. Soc. Rev., 2001, **30**, 168–176) The massive explosion arising from the death of a very large star, greater than eight times that of the Sun, is responsible for the scattering of heavier elements such as iron and gold synthesized within the progenitor star.

> The vast gas cloud Sagittarius B, at 150 light years across and close to the center of our Milky Way, is said to contain a billion, billion, billion liters of pure alcohol (ethanol), although you would need to mop up a huge area to collect enough for even a small glass of wine, since the molecules are so far apart.

Whereas some clouds of gasses are relatively quiescent, many others are constantly activated by shock waves or radiation to generate the process of accretion leading to star formation. The stars eventually die and blast off outer layers of gas to recycle the process. The result is that large numbers of gaseous entities, mostly unique in their precise composition and structure, display wonderful shapes and colors for us to photograph here on Earth. These are the nebulae, from the Latin word for "cloud." In many cases false colors are useful to present chemical or structural diversity or even unavoidable to allow them to be presented to our Earthly eyes that can see no more than a narrow range of the electromagnetic spectrum.

There are three main types of nebulae: emission nebulae, which give out light due to the activity of hot ionized gas; reflection nebulae, which reflect radiation from nearby sources; and absorption or dark nebulae, whose clouds have no source of emission and so block the light from anything glowing from behind. Several classes and subclasses of nebulae are described, including diffuse nebulae, planetary nebulae, reflection nebulae, supernova remnants, and dark nebulae. A planetary nebula is neither a planet nor a true nebula. It is a circle of gasses resulting from a star's blast toward the end of its life as it casts off its outer layer to leave a dense core. Diffuse nebulae are the beauty queens of space, with their wide-ranging pallete of colors and often bizarre shapes. Many are close to beginning star formation from the mass of gas available.

> Planetary nebulae were so labeled by William Herschel because they appeared, to him, somewhat similar to the shape and colored hues of the planets Uranus and Neptune.

M42 is a huge gas cloud, an emission nebula, in the sword of Orion just visible to the unaided eye and among the first nebulae to be identified. The spectacular image (see Fig. 6.16) illustrates dramatically the complexity of the region. The bright center is a hollow blasted out by Theta-1 Orionis, which actually hosts four of the nebula's largest hot stars, known as the Trapezium, and whose ultraviolet radiation also ionizes surrounding clouds. A mini nebula within the system, labeled M43, has just one massive star to blast its local gasses with radiation. The whole area, the size of four full Moons, is a vast star-forming region and, although just about visible with the naked eye, is quite apparent through a small telescope.

Location: The constellation of Orion, Orion's sword
Distance: 1,500 light years
Telescope: Hubble space telescope and European Southern Observatory's
     2.2-m telescope at La Silla
Camera: Advanced camera for surveys
Image: Part of a mosaic of 520 five-color Hubble images combined with
     ground-based photographs.

Monoceros was given its name because it was the 838th variable star found in the constellation of the same name. It is a reflection nebula. A short burst of light from the red supergiant about 2 years prior to image capture (2004) illuminated the surrounding, ever-expanding interstellar gas cloud that reflected the light back into space (known as a light echo), enabling us to see it. Maybe the reflecting dust originated from previous outbursts but was invisible until the light echo revealed its presence. As the light travels further through the clouds it reveals ever-changing patterns of dust (Fig. 6.17).

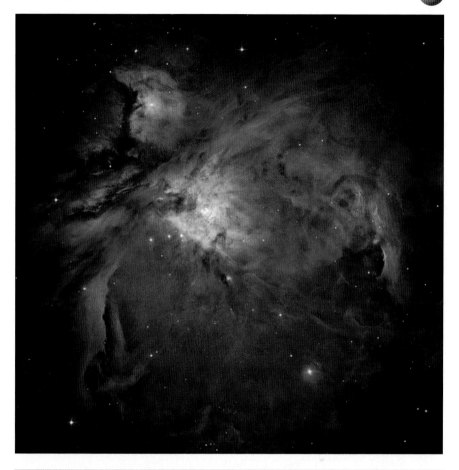

Fig. 6.16. A "close-up" of the intense star-forming region of the Orion Nebula, M42, or NGC 1976 (credit: NASA/ESA/ Hubble Heritage Team/STSci/AURA).

Location: Constellation of Monoceros
Distance: 20,000 light years
Telescope: Hubble space telescope
Camera: Advanced camera for surveys, Wide Field Camera
Image: The color image is composed of exposures through blue (5,250 s), green (1,050 s), and near infrared (1,050 s) filters.

A.D. Thackeray first observed the following example of a dark nebula in 1950, although Dutch–American astronomer Bart Bok, in 1947, drew attention to the existence of such globules that are among the smallest of

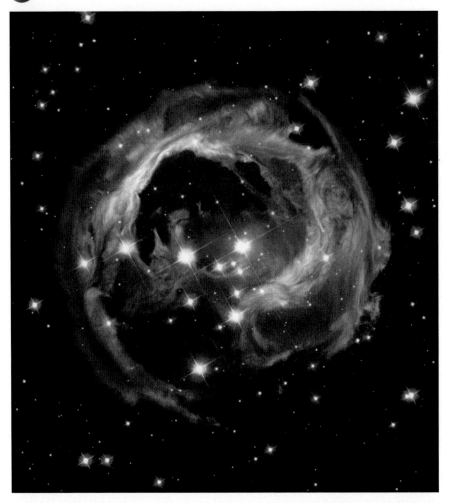

Fig. 6.17. A light echo reveals the presence of local interstellar dust around V838 Monoceros (credit: NASA/ESA/Hubble Heritage Team/STSci/AURA).

the dark nebulae. The star-forming region of IC 2944 is filled with gas and dust illuminated by a loose cluster of massive hot stars that provides a backdrop to reveal the opaque nature of the globules (Fig. 6.18). Ultraviolet radiation from the surrounding stars streams against and heats the globules, leading to eventual destruction of the dark, dense, opaque clouds of dust. Some of the absorbed light is reradiated as infrared radiation.

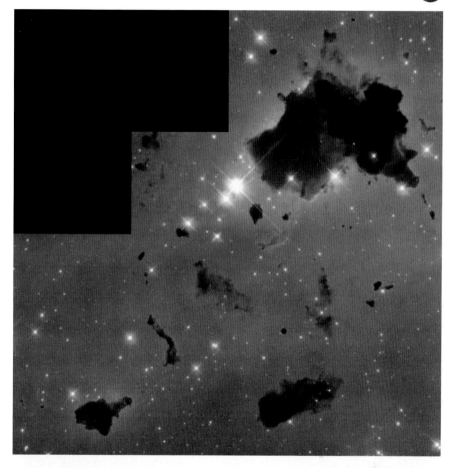

Fig. 6.18. Opaque dust clouds or "globules" in IC 2944 (Thackeray's Globules) (credit: NASA/ESA/Hubble Heritage Team/STSci/AURA).

The Horse Head Nebula in Orion (see Fig. 1.10 in Chap. 1) is another well-known example of a dark or absorption nebula.

Location: The southern constellation of Centaurus
Distance: About 5,900 light years
Telescope: Hubble space telescope
Camera: Wide field and planetary camera 2
Image: A composite based on the original hydrogen-emission image from red, green, blue, and H-alpha filters

Exotic Stuff

Mysterious matter resides in the cosmos! As little as around 5% of the universe is visible to modern telescopes and detectors of all kinds. The remainder has been dubbed "dark matter", stuff that has been predicted in an attempt to account for the composition of space according to current models. The search is on for clues to, or even evidence of, "dark existence." A recent (1992) discovery points to the possibility that dark matter is not evenly distributed but concentrated in and around small groups of galaxies. X-ray pictures of a small cluster of galaxies, the NGC 2300 group (see Fig. 6.19), show that it is surrounded by a huge cloud of hot (10 million Kelvin) gas 1.3 million light years in diameter. But it should be larger or even dissipated, unless it was held together by the gravity of an immense mass of unknown identity. The colorful photograph certainly could be a pointer to an important piece of the jigsaw puzzle.

Also hoping for clues to, or even evidence of, dark matter is the Large Hadron Collider on the Swiss/French border that has now been switched on.

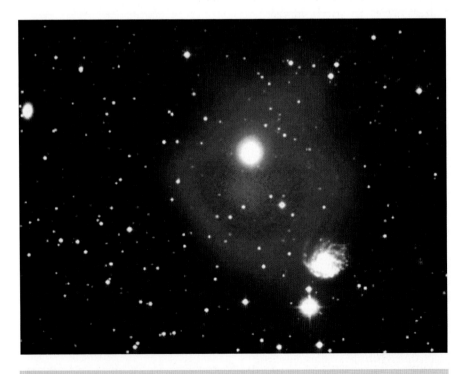

Fig. 6.19. Evidence for dark matter in NGC 2300? (credit: NASA/ROSAT/STSci).

The massive project is associated with truly extreme specifications and expectations. Protons will be accelerated around a 27-km ring deep underground to within a tiny percentage of the speed of light, 99.9999991%, to then encounter a head-on collision possibly creating particles never seen before or even mini black holes. Many of these particles will exist for less than a trillionth of a second, and some will be able to sneak unseen through many meters of detector.

At 1.9K the temperature of the ring will be colder than intergalactic space, necessary for the operation of superconducting magnets made of niobium-titanium to control the acceleration and the pathway of the protons at such a high energy. In contrast, the temperature generated at the point of collision will be thousands of times the temperature at the center of the Sun. One expectation/hope is to discover the Higgs boson postulated to be responsible for the mass of particles. Confirmation of the super symmetry theory (SUSY) that postulates the existence of a heavy super partner for all known particles might be forthcoming. The lightest of the proposed super symmetric particles is a candidate for dark matter.

Location: In the direction of the constellation Cepheus
Distance: 150 million light years
Telescope: ROSAT (Roentgen Satellite) X-ray observatory satellite, Germany, NASA and UK
Camera (detector): Positive sensitive proportional counter
Image: False color to present the X-ray image

Black holes are the stuff of fantasy – cannot see inside them, not absolutely and finally sure they even exist, they swallow stars, turn you into spaghetti if you fall into them, and are even heavier than you can imagine. Confirmation of the existence of black holes is more likely, however, following a study of the motions of 28 stars orbiting our galactic center in the direction of Sagittarius. It is estimated that the super massive black hole is around four million times as massive as our Sun. It is even proposed that there may be many black holes in our own galaxy hidden within strange star clusters emanating from the many smaller galaxies that smashed together to form the Milky Way (Fig. 6.20).

Stars around the size of our Sun will eventually cease to generate energy through nuclear fusion and collapse to form a white dwarf of incredible density. One lump the size of a sugar cube would weigh tens of tons! The white dwarf, retaining huge quantities of heat, would slowly cool and, given enough time – maybe more time than that of the universe so far – become a frozen ball of inactivity. Stars much larger than our Sun, as they died, would succumb to their greater gravity to squash protons and electrons together to form neutrons. Our lump of sugar-sized material of the neutron star would then weigh thousands of millions of tons.

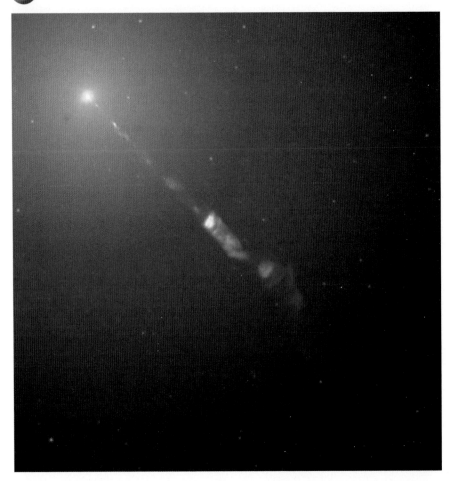

Fig. 6.20. Black hole-powered stream of particles from the center of M87 (NGC 4486) (credit: NASA/Hubble Heritage/ STSci/AURA).

The final destiny of very large stars would not stop there, and their further collapse, following cessation of all nuclear reactions, would result in the formation of a black hole. So great would be the density that the escape velocity would be too great to allow even light to escape. If light cannot escape from a black hole it cannot be seen, it can only be inferred from its effect on other bodies. Black holes cannot always cope with the rate of inflowing debris and squirt out the excess at enormous speeds along the central axis of the rotating accretion disk, possibly due to the interaction of ionized material with powerful magnetic fields. At the center of M87 is a super massive black hole of about two billion solar masses that ejects a beam of subatomic particles at almost speed of light 100,000 light years long.

Location: Constellation of Virgo
Distance: 50 million light years
Telescope: Hubble space telescope
Camera: Wide field and planetary camera 2
Image: The color image, contrasting the blue of the jet with the yellow background of a myriad stars, was created through a combination of exposures of ultraviolet, blue, green, and infrared light.

Our Sun

We cannot leave the subject of nebulae without showcasing our very own star, which may itself be the origin of a nebula in a few billion years time. There have been several satellites dedicated to the study of the Sun's structure and solar emissions, since they clearly have important consequences for Earth, its inhabitants, and future manned space travel. The *Yohkoh* (Sunbeam) carrying spectrometers and telescopes was launched in1991 from the Kagoshima Space Center in southern Japan to observe the energetic phenomena taking place on the Sun, specifically X-ray and gamma ray emissions. The *RHESSI* mission, Ramaty High Energy Solar Spectroscopic Imager, was launched in 2002 from the Kennedy Space Center (Cape Canaveral) with the ability to obtain images and high resolution spectra over a wide range of X and gamma rays, with the purpose of studying solar flares. NASA/ESA's *Ulysses* spacecraft is the first to orbit the Sun over its poles to monitor the occurrence, strength, and polarity of magnetic fields and characterize the heliosphere as a function of solar latitude (see Fig. 6.21). *Ulysses* was launched from the shuttle Discovery

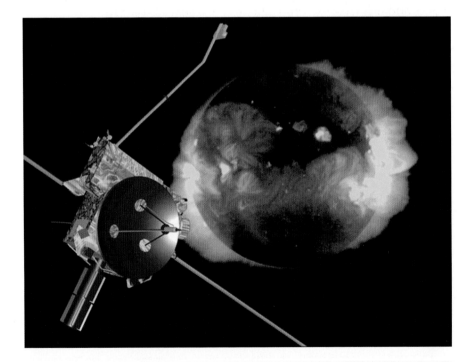

Fig. 6.21. Composite picture, *Ulysses* first perihelion – March 1995 (credit: NASA/JPL).

in 1990 toward Jupiter for a gravity assisted fly-by to accelerate the craft out of the ecliptic plane to high latitudes.

STEREO (or the Solar TErestrial RElations Observatory) is a 2-year mission to study coronal mass ejections (CMEs) and solar wind using two almost identical craft, hence stereoscopic measurements. The twin observatories were launched from Cape Canaveral in October 2006. Lunar gravity was employed to place one ahead of the Earth's orbit and one trailing. *SOHO* (or SOlar Heliospheric Observatory) is an ESA/NASA joint venture, the craft being built for ESA by a European multinational team led by the company now called EADS Astrium. *SOHO* was designed to study the internal structure of the Sun, the corona, and the solar wind out to a distance ten times beyond Earth's orbit. Launched from Cape Canaveral in1995 and parked in the first Lagrangian point, L1, it is part of an international program to which Japan, Russia, Sweden, and Denmark also contribute satellites to monitor the Sun and solar effects.

The European *Cluster* mission to study small-scale structures of the magnetosphere in three dimensions finally launched from the Russian Baikonur Cosmodrome in 2000 after a disastrous explosion of the Ariane 501 from French Guiana in 1996. *SOHO* and *Cluster* are both part of ESA's Cornerstone science project. Figures 6.22, 6.23, and 6.24 all present images and information captured in false colors, of course.

References are given in the Appendix of this book for more details concerning the missions or images, many of which are movies.

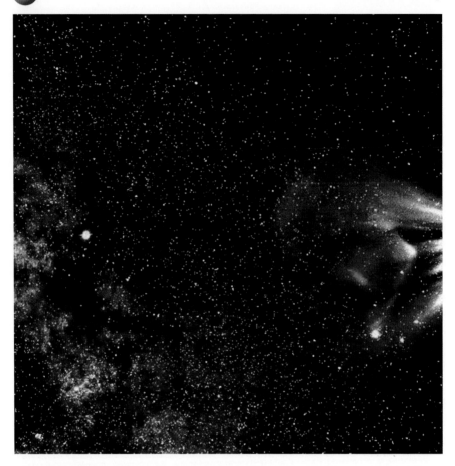

Fig. 6.22. *STEREO* image showing a coronal mass ejection (CME), the Milky Way, Jupiter (*left*), and Antares (*lower right*) (credit: NASA).

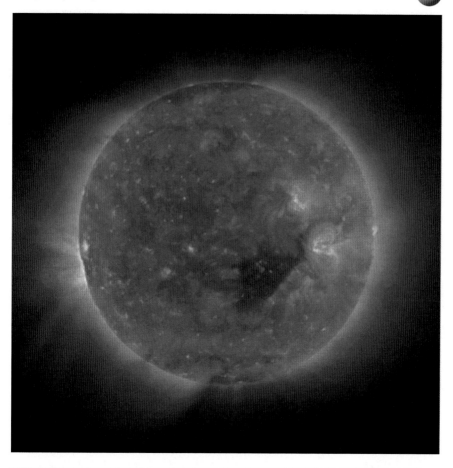

Fig. 6.23. *STEREO*'s SECCHI/EUVI telescope at 195 A (credit: NASA).

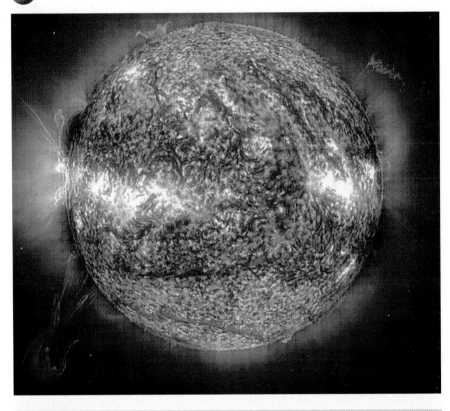

Fig. 6.24. A huge prominence, ten times the size of Earth, erupts from the Sun's surface (credit: *SOHO* consortium/EIT/ ESA/NASA).

Amateur Gems

Professional images of deep sky objects employing huge, expensive telescopes in optimum locations are absolutely stunning. However, amateur astronomers also have much to offer, especially as the price of sophisticated equipment incorporating advanced technology drifts into the range of affordability. Throughout the world a huge number of local astronomical societies harbor a wealth of talent. Some societies concentrate on discussions, others on naked eye or binocular observation, and still others on the capture of images of the highest quality. Fay Saunders of the Orpington Astronomical Society in England, within a very short time of acquiring reasonably inexpensive equipment and with much dedication, produced an image of the Rosette Nebula to rival some of the best available (Fig. 6.25).

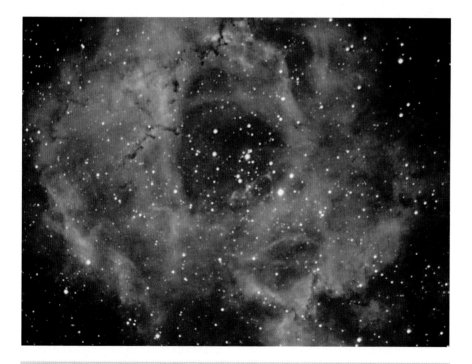

Fig. 6.25. The Rosette Nebula in Monoceros (credit: Fay Saunders, Orpington Astronomical Society).

Fay Saunders writes:

"The Rosette 'complex,' a vast cloud of dust and gas with a central open cluster, is known as NGC2237. It is a very active region of star formation and is the finest example of an HII area in the northern sky. It is 100 light years across and 3,000 light years away. The nebula is made up of NGC2238 and NGC2239, discovered by John Herschel, and NGC2244, which is the central open cluster discovered by John Flamsteed. This cluster is the result of star formation within the nebula and its stellar winds are clearing a hole in the nebula center. The cluster, which formed around four million years ago, emits an ultraviolet light that causes the surrounding nebula to glow.

"This was the first nebula I had imaged and processed using RGB and Hα filters. I was amazed at its beauty and interesting complex structure, resembling a huge flower, with the filaments of dark dust running through the nebula's gasses."

Distance: 2,600 light years
Diameter: 130 light years
Equipment: Starlight Xpress SX M7, Ha, Cls filters, Skywatcher Pro ED80. Skywaycher Pro HEQ5 mount
Guided: WO ZS66 0.8FR, PHD (guiding software)
Details: L = 18 × 300 s RGB 5 × 300 s each
Processed: AstroArt, Maxim DL, PS CS3
Imaged: Chislehurst, Kent, UK, 090208 18.23

A solar eclipse is a wonderful sight, especially if you are lucky and skillful enough to catch images just at the right time.

Camera: Olympus E55 DSLR, 1,000 mm mirror lens
Settings: 1/500 s at f8, iso 400
Location: Turkey (near Antalya), 29/03/2006, 10:56 a.m.

Jim Mehta writes:

"We were by the seaside in Turkey and were lucky enough to have a beautiful sunny day for the event. In fact it was a scorcher. Quite exciting waiting for the event, there were about 100 people in the group all with scopes, so quite a buzz as everyone was setting up. I was traveling light and only brought my camera and lens, a lightweight photographic tripod, and some small binoculars with homemade solar filters as I was more interested in experiencing the eclipse than taking pictures. There was great excitement as the Moon made first contact; a cheer went up and slowly twilight crept up. Using the binoculars I was able to observe the Sun as the Moon slowly covered it.

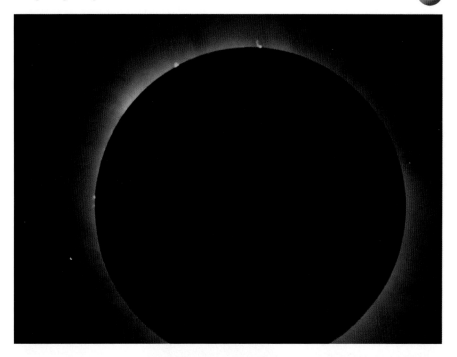

Fig. 6.26. Fire from the shadow (credit: Jim Mehta, Orpington Astronomical Society).

One interesting feature was how the shadows changed. They were no longer distinct but blurry, very weird! It became cold and very still. The frogs that had been making a racket from the fields opposite shut up and finally totality arrived. The darkness rushed towards us from the sea as the light faded. What surprised me was how much light was still available (more than on a bright moonlit evening). I started taking pictures. I had preset the focus to infinity and took a sequence of pictures using different shutter speeds, that took maybe a minute, then I just observed. I saw Mercury and Venus high in the sky. Then as quickly as it had begun it was over." (Fig. 6.26)

It must be emphasized that WHEN VIEWING THE ECLIPSE DIRECTLY BINOCULARS AND TELESCOPES MUST BE APPROPRIATELY SHIELDED WITH FILTERS UNTIL THE SUN IS COMPLETELY IN SHADOW (Fig. 6.27).

M27, the Dumbbell Nebula, is a planetary nebula in the constellation of Vulpecula 1,200 light years away and 2.5 light years across. It is a gaseous emission nebula formed as a star runs out of fuel to "burn" and

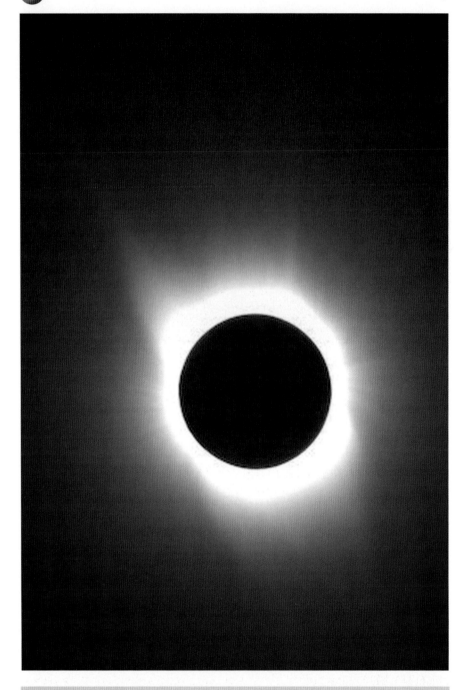

Fig. 6.27. The solar corona as visible during an eclipse on August 1, 2008, over the Gobi desert (credit: Joe Cali. Visit Cali's website for more stunning images, http://www.joecali.net).

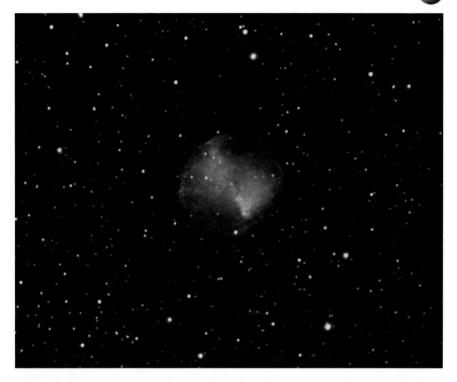

Fig. 6.28. M27, the Dumbbell Nebula (credit: Mike McRoberts, Orpington Astronomical Society).

blows the star's outer layers into space. The gas is illuminated owing to an intense blast of ultraviolet radiation from the dying star exciting the atoms of the nebula. M27 was the first planetary nebula discovered, by Charles Messier in 1764 (Fig. 6.28).

There are some beautifully wispy nebulae, such as IC 1805. The Heart Nebula (see Fig. 6.29) is a cluster of stars along with dust clouds silhouetted against glowing gas. A young emission nebula around 1.5 million years old, it is about 7,500 light years away. Figure 6.30 shows the Veil Nebula, formed from a star that exploded over 5,000 years ago that would have been seen as a "star" many times brighter than Venus at its brightest. It lies 1,400 light years away toward the constellation of Cygnus.

And possibly one of the best known deep sky images is M16 (see Fig. 6.31). M16 is part of the Eagle Nebula, a massive star-forming region 7,000 light years away in the constellation of Serpens. The image is composed of three colors – red, green, and blue – to represent emissions from singly ionized sulfur atoms, hydrogen, and doubly ionized oxygen atoms.

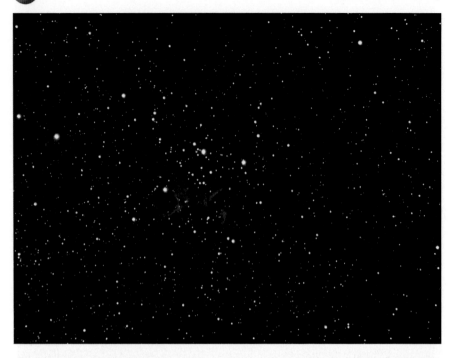

Fig. 6.29. The Heart Nebula (credit: Mike McRoberts, Orpington Astronomical Society).

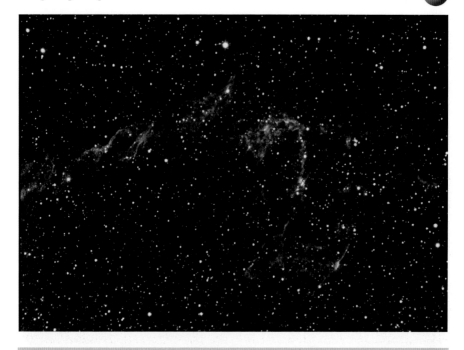

Fig. 6.30. The Veil Nebula (credit: Mike McRoberts, Orpington Astronomical Society).

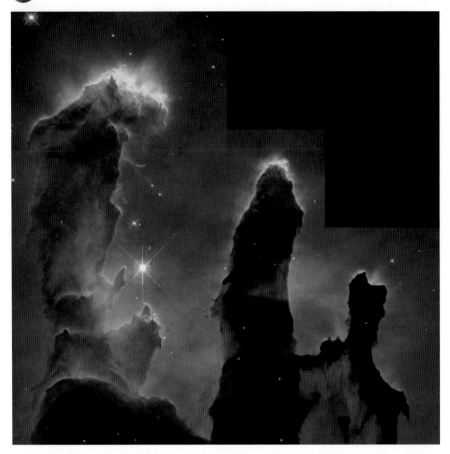

Fig. 6.31. Pillars of creation (credit: NASA/Jeff Hester and Paul Scowen at Arizona State University).

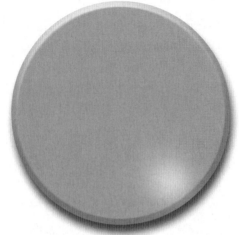

CHAPTER SEVEN

Solar System Colors

The *Cassini* mission pictures of Saturn are amazingly detailed and contain dazzling awe-inspiring colors, especially the colors of the rings. But wait a moment! Are they really the true colors that you would see if you were as close as the *Cassini* craft? Maybe not! That begs the question as to whether any of the astronomical colors we see are for real. What is real color and what is false?

Real color could be stated as that which we see when close enough and without assistance from optical instruments of any sort, except spectacles. However, even that cannot be correct. As described in a previous chapter, there is great variation between people as to exactly what they see, or to be more precise, how the brain interprets the electrical pulses generated by light falling on the eye's rods and cones. Such is the variation that it is likely no one can guarantee that the hues they see are the same as those someone else might see, which means there is no such thing as real color, if defined as the colors we see. Take an extreme example for easy illustration.

During a country walk on a sunny day you might come across a stunning vista (*see* Fig. 7.1), a vast area of poppies generously allowed by the farmer to exist for the pleasure of those venturing along a pathway through the valley. You take a digital picture and place it on your computer screen at home to view with someone who has some color blindness. What did that person see? What did you see? Imagine you then moved

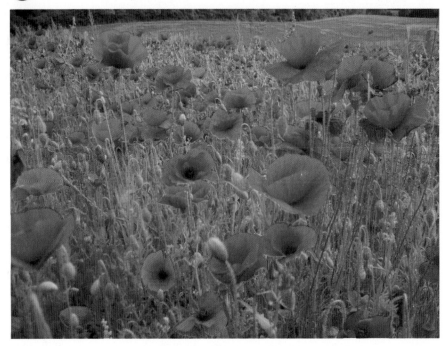

Fig. 7.1. A country poppy field.

the color filter slider to turn off the red. The red color blind person would then be unable to discern any difference between the two pictures. He or she would only see black poppies. (*see* Fig. 7.2). Black poppies!

Although tiny variations of color perception between people might not be noticed, real color to one may be different from real color to another. Whereas lack of "red cones" is only prevalent in about 6 to 8% of humans, it is 100% in some lemurs. In fact, many mammals have just two kinds of the light-sensitive pigments, called opsins, and therefore are sensitive to short and middle wavelength light – blues and greens.

Most birds are better off in terms of color vision. It is a dead giveaway that the males are highly colored to attract the female. Some birds, bees, and insects are even able to see ultraviolet, and the rattlesnake can sense infrared, although it does not use its eyes but a detector in a hole or pit in front of each eye. Hence the alternative name of pit vipers.

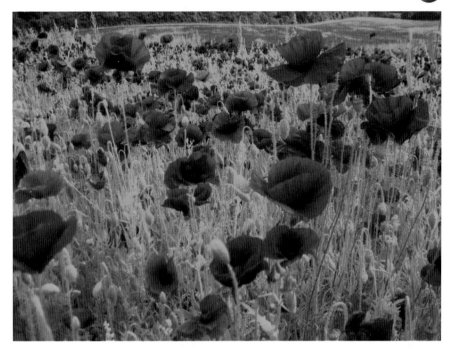

Fig. 7.2. Black poppies!

Nocturnal creatures have less need for color; owl monkeys, at the extreme, are monochromatic and only see black, white, and grays. At the other end of the scale many animals, including some spiders, birds, reptiles, and fish, are tetrachromats who have four types of eye color receptors, and their perception of color is enhanced. It is said that maybe half of all women, but very few (if any) men, are tetrachromats.

For astronomical imaging, however, two categories can be defined – those that attempt to capture or reproduce what an "average" person might see and those that are deliberately falsified to represent differences in physical or chemical properties. (Of course, there are other categories not relevant to this work, such as those that include artistic impressions of the unknown, to illustrate hypotheses or for fictional material.)

So, what color are our planets and what can be discovered and illustrated through the production of false color images?

The Planets

Mercury

Our innermost planet is small, even smaller than some of the larger moons in the Solar System. At 4,879 km in diameter it is just a bit larger than our own Moon, which is 3,476 km in diameter, and smaller than the other large moons, Ganymede (moon of Jupiter), 5,262 km, and Titan (moon of Saturn), 5,150 km. It is officially now the smallest planet, since poor Pluto, at 2,390 km, was downgraded to a dwarf planet or Kuiper Belt object. Its highly elliptical orbit takes Mercury so close to the Sun, 46 million km at perihelion, that very little in the way of atmosphere can exist apart, possibly, from temporary gassing as cosmic debris thuds into its roasting hot or freezing cold surface or hydrogen and helium are temporarily captured from the solar wind.

Mariner 10 was the last of an illustrious series of space probes. *Mariner 2* was the first probe to visit another planet, Venus; *Mariner 4* gave us close views of Mars, and *Mariner 9* was the first to orbit another planet, Mars. *Mariner 10* flew by Mercury three times, showing, with its two visible-light filters in the wide angle camera, a planet that was very gray and peppered with craters similar to the Moon.

A new mission, *ME*rcury *S*urface, *S*pace *EN*vironment, *GE*ochemistry and *R*anging spacecraft, *Messenger,* was launched on March 8, 2004, from Cape Canaveral in Florida. After a complex flight path using Earth, Venus, and Mercury to reduce its speed it will settle into Mercury's orbit on March 18, 2011. *Messenger* made a close flyby of Mercury on January 14, 2008, and sent back over 1,200 stunning black and white images. A second close approach occurred in October 2008, and the next is planned for 2009.

The mass of craters, clearly, is similar to the Moon, but there are important differences. These and other facts will be accumulated and fully analyzed as the mission unfolds, but one picture currently available is the fascinating false color picture created by converting three images taken at 1,000, 700, and 430 nm (infrared, far red, and violet) to red, green, and blue before combining them to produce a false color photograph (*see* Fig. 7.3). Color differences on the surface of Mercury are therefore accentuated. Although the color differences are rather pale, some bright spots with a bluish hue indicate that they are relatively recent impacts. A more vivid revelation of the subtle color of the surface has been made by increasing the saturation in an image enhancement computer program where the blue is bright and the surrounding surface quite red, possibly an indication of the presence of iron. We must be patient, though, before drawing too many conclusions.

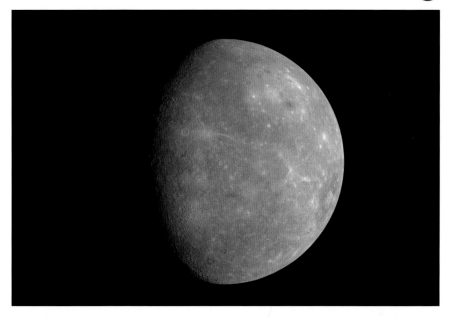

Fig. 7.3. Mercury in color (Credit: NASA/JPL/John Hopkins Applied Physics Laboratory).

Cape Canaveral was renamed Cape Kennedy in 1963 as a memorial to the assassinated President John F. Kennedy, who was a great supporter of the space program. The name reverted to Cape Canaveral in 1973 to recognize that the name had actually been in existence for 400 years.

Venus

Venus is familiar to most as the Evening Star or Morning Star, depending on its position relative to the Sun at the time of observation. Of course, it is not a star but our second planet from the Sun and by far the brightest, with a blazing maximum magnitude of −4.7. In size it is Earth's sister, 12,104 km in diameter compared to our own 12,756 km. In many other respects it is so very different, our "evil twin." Often referred to as global warming gone mad it is shrouded with a thick and dense layer of cloud, not any old water cloud as on Earth but mostly carbon dioxide and some sulfuric acid. Only traces of water and a few other gases exist. The pressures and temperatures of the atmosphere are what might be expected from such a lethal mix, 92 times the pressure of Earth's and an average of 460°C. It would be fun to see what happens, from a safe distance, as a can of fizzy drink were dropped into the atmosphere, to see it collapse as if squashed in a vice! Slimline drink indeed!

Although telescope views of Mercury have shown great details of its surface, there has been no such luck with Venus. As amateur astronomers are all too aware, the thick, dense clouds prevent all but a fluffy white ball appearing in the eyepiece. Moonlike phases are great to observe and photograph, but one moves on rather quickly from that. How wonderful it is, then, that scientists have come to the rescue. Radar maps have been made of the surface that reveal the existence of hills, volcanoes of all sorts, and, at 6,000 km long, Baltis Vallis, the longest lava channel in the Solar System. Impact craters also exist, but the dense atmosphere prevents small meteoroids, of a size that would become meteorites on Mercury, from reaching the surface.

So the real or true color of Venus is white or off-white as recorded by *Messenger*, Hubble, and many astrophotographers, owing to the thick atmosphere hiding the surface (*see* Fig. 7.4). It is well-known now that the surface is a sweltering 460°, so the false color, computer-generated images of the surface constructed by combining Magellan's radar images, and the color images of the Soviet *Venera* missions 13 and 14, delightfully

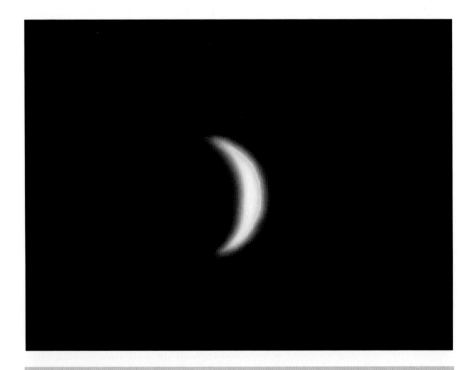

Fig. 7.4. Cloudy Venus.

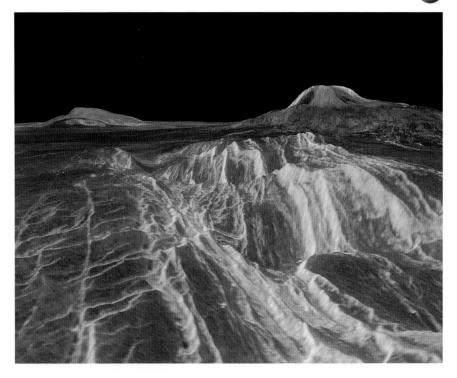

Fig. 7.5. Venus's fiery surface (Credit: NASA/JPL).

present the expected fiery furnace picture (*see* Fig. 7.5). The volcano Gula Mons is on the horizon to the right. The Galileo Orbiter, on its way to study Jupiter in 1990, employed false color imagery to bring out the fine structure of Venus's cloud markings (*see* Fig. 7.6). One picture can represent an amazing amount of time spent, the scientists' involvement, and technology. One such is from the Magellan Imaging Radar. A mosaic of images, with contributions from the Earth-based Arecibo radar, *Venera* spacecraft, and *Pioneer* missions were combined and color-coded to represent elevation in this wonderful hemispheric view (*see* Fig. 7.7).

Little can beat the excitement of witnessing that rare occasion of a transit of Venus across the Sun. Not only is it a once-in-a-lifetime spectacle but the stunning color contrast of the golden Sun in the heat of the day hosting a very black spot is awesome indeed. Figure 7.8 shows a gathering at Sir Patrick Moore's house for the 2004 transit of Venus.

Earth

We know Earth has lots of blue, and we walk all over the place surrounded by green, so not a lot to say, except…what do we look like to others? Not

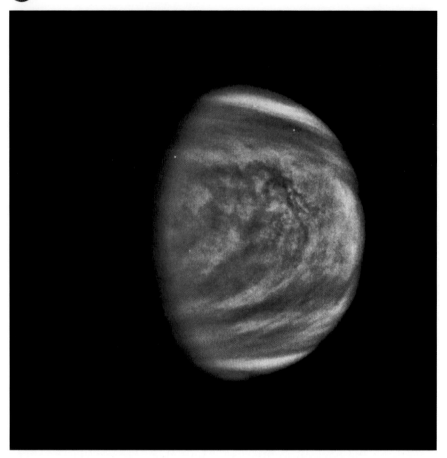

Fig. 7.6. Venus's cloud patterns (Credit: NASA/JPL).

other people but what do we look like when viewed from other cosmic bodies? Best not to go on too much about images sent from our local artificial satellites, such as the Shuttle Radar Topography Mission (SRTM) shown in Fig. 7.9, a radar image of craters in Tanzania; that would be an opus in itself. (One compilation of stunning images is contained in the book entitled *Orbit* (*see* the Appendix of this book). Images of Earth are sometimes taken to calibrate or test equipment aboard a leaving spacecraft.

Messenger, on its 2005 flyby, turned the dual imaging system's wide angle cameras towards Earth to produce images designed to identify minerals on a planet's surface. From an altitude of around 100,000 km two particular images were captured, varying in only one wavelength band. The first, using filters peaking in sensitivity at 630, 560, and 480 nm, approximates to what the human eye can see. This natural color helps to distinguish

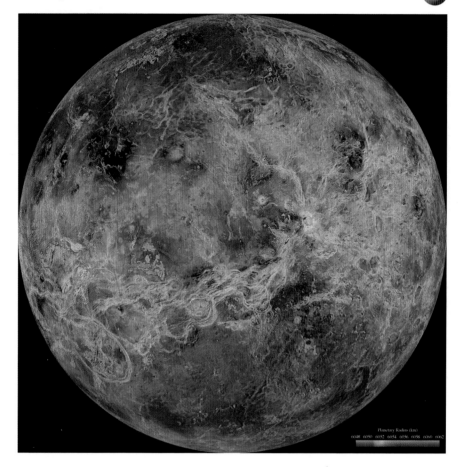

Fig. 7.7. The result of a decade of radar investigations (Credit: NASA/JPL/U.S. Geological Survey).

materials with distinct visible color differences, such as ilmenite and volcanic glasses. The second image, using 750, 630, and 560 nm concentrates more on revealing green vegetation represented by red on the image (*see* Fig. 7.10). Selection of wavelengths optimizes for pictures to take into account sharp (near-infrared) and scattered light (blue). The full range for *Messenger* cameras covers 400–1,000 nm.

Mars

So much has been discovered, photographed, and written about our most explored planet that it is difficult to choose just a few images to illustrate

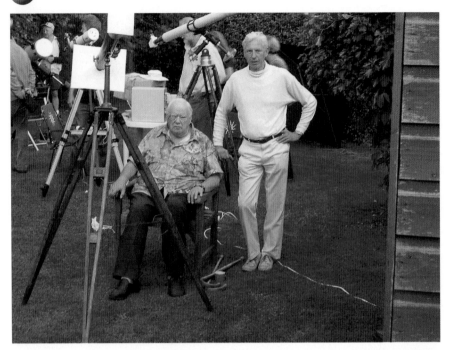

Fig. 7.8. The author with Sir Patrick during the transit of Venus.

its awe and color. Much of its fascination has revolved around the possibility of the existence of life. The standard content of many science fiction stories are little green Martian men with big heads and bulbous eyes. The misinterpretation of surface features as canals did nothing to dispel the notion, especially promoted by the American businessman Percival Lowell (1855–1916), who established the Lowell Observatory in Arizona in 1894 to search for intelligent life on Mars.

Giovanni Virginio Schiaparelli (1835–1910), an Italian astronomer, first noted the association of the Perseid meteor events with the Comet Swift-Tuttle among his many discoveries. From 1877 his careful mapping of the surface of Mars showed straight lines he called "channels," or canali, that were referred to thereafter as canals and hence attributed to intelligent life. Not until the Mariner probes of the 1960s was the myth finally laid to rest.

Considering how barren and "airless" Mars is, it is surprisingly sometimes referred to as the most hospitable planet in the Solar System after Earth. At around 6,780 km in diameter it is almost half that of Earth, with a volume of about one-sixth. It has very little atmosphere, only about 1% of Earth's, but what is there is comprised mostly of carbon dioxide,

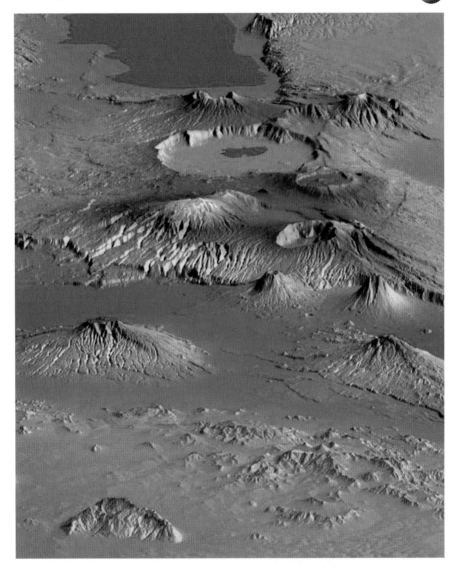

Fig. 7.9. Crater Highlands, Tanzania (Credit: NASA/JPL/National Geospatial-Intelligence Agency [NGA]).

with some nitrogen and argon and traces of others gases such as oxygen and carbon monoxide. Although thin, the atmosphere still hosts serious weather events, such as dust devils, clouds, and possibly snow. Even modest telescopes with a webcam can record, infuriatingly, a massive dust blanket preventing further photography until settled.

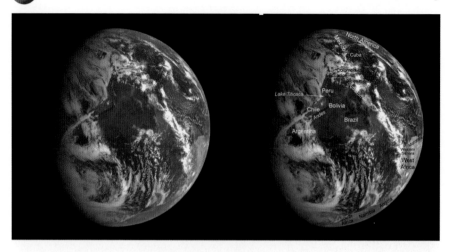

Fig. 7.10. *Messenger's* dual images of Earth (Credit: NASA/ John Hopkins University Applied Physics Laboratory/Carnegie Institute Of Washington).

Over the last four decades there have been dozens of missions to Mars launched by many countries. More than half have failed, and some did not even make it off the launch pad. The purpose of the missions was to glean information about the geology, history, and material content, in particular, the presence of water and organic molecules to give clues to the present and past existence of life (exobiology). Information gathered will be vital for the planning of manned spaceflights – which might become a reality at some time in the future – to the inhospitable planet.

A major positive for us on Earth is the fantastic collection of true and false-color breathtaking images from international space organizations that allow us to see the surface close up. One could almost imagine planting a deckchair in the warm and pleasant-looking Mars-scape. Plenty of beach, but where's the water?

An early, stunning representative color image of the largest volcano in the Solar System, Olympus Mons, was captured by the *Mars Global Surveyor,* launched in October 1998 (*see* Fig. 7.11). We are now showered with amazing high-quality pictures from the two rovers, Spirit and Opportunity, placed on opposite sides of the planet in January 2004. Also amazing is that they have continued to function effectively for many, many times longer than expected, producing thousands of useful and colorful images.

From November 6–9 in 2007, Spirit perched near the western edge of "home plate"; its panoramic camera, Pancam, combined three images

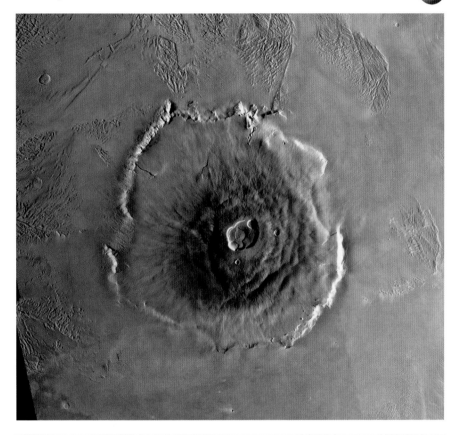

Fig. 7.11. Olympus Mons from orbit (Credit: NASA/JPL/Caltech).

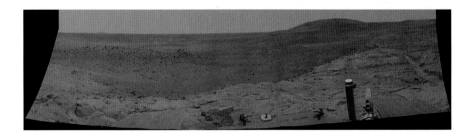

Fig. 7.12. Gusev Crater from Spirit, true color (Credit: NASA/JPL-Caltech/Cornell University).

of 735, 535, and 432 nm to produce a true color view of Gusev Crater (*see* Fig. 7.12) and a false-color stretch (*see* Fig. 7.13) to bring out subtle color differences.

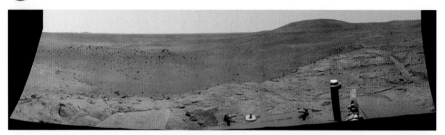

Fig. 7.13. Gusev Crater from Spirit, false color (Credit: NASA/ JPL-Caltech/Cornell University).

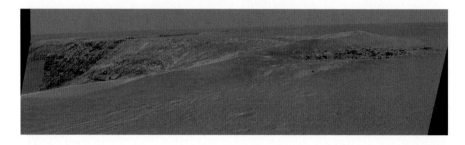

Fig. 7.14. Cabo Frio from Opportunity in "Duck Bay" (Credit: NASA/JPL/Cornell).

During its 952nd sol, or Martian day, Opportunity's Pancam produced an approximately true color rendition (*see* Fig. 7.14) of the promontory Cabo Frio in Victoria Crater, from which it is easy to understand Mars's nickname as the Red Planet.

While gazing in awe at the rover's achievements, other stunning missions must be acknowledged. The European Space Agency's *Mars Express Orbiter* has accumulated valuable geological information. The combination of a topographical map with radar scans of the subsurface presented in false color, of course, contributes to the study of a mysterious formation dubbed the Medusae Fossae (*see* Fig. 7.15).

The magic continues as the *Phoenix's* robotic arm digs, and its Surface Stereo Imager photographs operations by another lander (*see* Fig. 7.16).

On the 111th Martian day, sol, the Surface Stereo Imager captured the image in Fig. 7.17 of the spacecraft's crumpled heat shield and bounce mark on September 16, 2008.

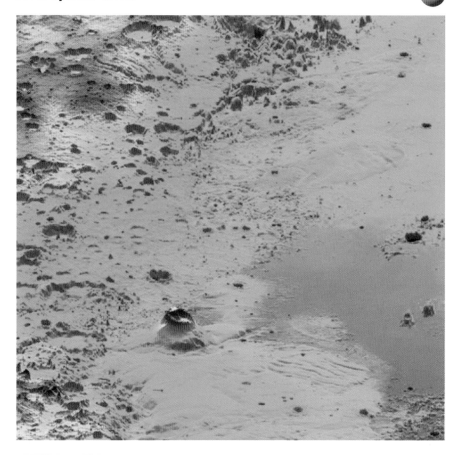

Fig. 7.15. The Medusae Fossae formation may represent some of the youngest deposits on the planet (Credit: NASA/JPL-Caltech/ESA/Italian Space Agency/University of Rome/Smithsonian).

ESA's *Mars Express,* using the High Resolution Stereo Camera, has returned images of Echus Chasm, one of the largest water source regions on the Red Planet (*see* Fig. 7.18).

The dark material shows a network of light-colored, incised valleys that look similar to drainage networks known on Earth. It is still debated whether the valleys originate from precipitation, groundwater springs, or liquid or magma flows on the surface.

Many thousands of almost unbelievable images are accessible on NASA and ESA websites.

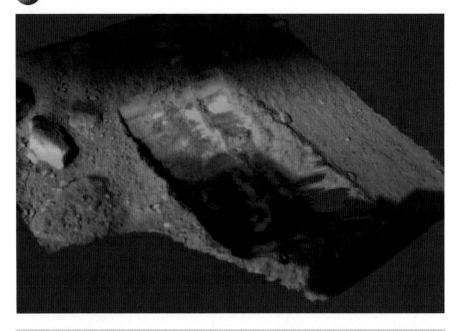

Fig. 7.16. Color-coded elevation map of the "Dodo-Gold-ilocks," 7–8 cm trench (Credit: NASA/JPL-Caltech/University of Arizona/Texas/NASA Ames Research Center).

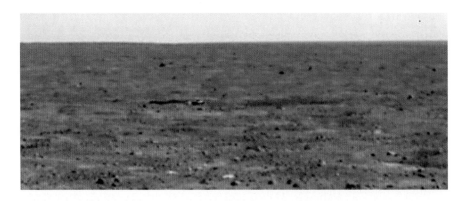

Fig. 7.17. Phoenix's crumpled heat shield (Credit: NASA/JPL-Caltech/University of Arizona/Texas A&M University).

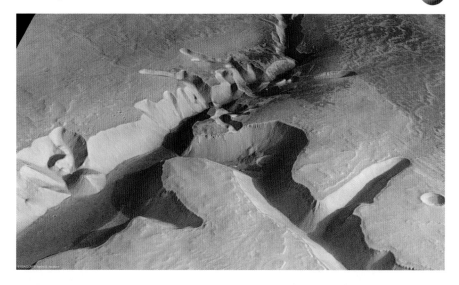

Fig. 7.18. Echus Chasm is the source region of Kasei Valles, which extends 3,000 km to the north (Credit: ESA/DLR/FU Berlin – G.Neukum).

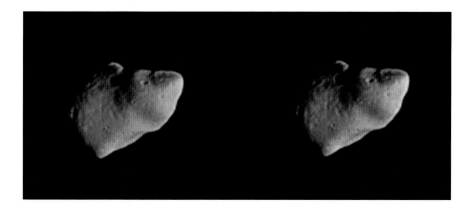

Fig. 7.19. Asteroid 951's (Gaspra's) colors, true (*left*) and false (Credit: NASA/JPL/US Geological Survey).

Asteroids

Stunning as the colors are for many of the major bodies of the Solar System, it must not be forgotten that there are probably millions if not billions of small and large rocks, many of which reside between the orbits of Mars and Jupiter – the Asteroid Belt. Figure 7.19 shows two color views of asteroid Gaspra, which is 19 km along its longest axis. The image was captured from the *Galileo* spacecraft (a mission to study the Jupiter system) on October 29, 1991. On the left is shown an approximate true color representation and the enhanced colors on the right show variations that may be due to surface composition or texture.

Plans are afoot to fly a spacecraft to an asteroid within a decade to drill for samples, which could contain pristine material formed at the same time as the Solar System. The European mission is called Marco Polo, with a major involvement of the space satellite manufacturers EADS Astrium. The European Space Agency, ESA, hopes to achieve the mission objectives around 2017 if the decision to go ahead is taken. NASA, in the United States, is working on a mission to put humans on an asteroid (and, hopefully, to bring them back again!).

> Marco Polo (1254–1324) was an Italian explorer who accompanied his father and uncle on many trading trips and returned from China with fantastic stories of the Chinese/Mongolian cultures and things Europeans had never heard of. Many disbelieved Marco's tales, but even on his death bed he refused to recant his tales.

Jupiter

Big is hardly the word for this gas giant! It is huge, with a diameter of around 150,000 km. In terms of mass or volume it could swallow all the other planets put together. Its composition of mainly hydrogen and helium could have allowed it to become another Sun if it were about eighty times more massive. To become a star a body would have to be at least 0.08 solar masses. Its magnetic field is 20,000 times greater than Earth's, and it is surrounded by an intense radiation belt of electrons and ions trapped in the magnetic field. The magnetic field is produced by rotation of the metallic liquid hydrogen formed under massive pressures deep within the Jovian atmosphere.

> The Sun is 1,047.56 Jupiter masses, so, $0.08 \times 1,047.56 = 83.80$ times more mass would be required for Jupiter to attain the minimum for star existence.

The rotation time of just under 10 h causes ferocious winds in Jupiter's upper atmosphere, leading to gaseous layers or belts visible even in modestly sized telescopes. The planet may be hostile, but it is at the same time beautiful, and the beauty lies in the differentiation of the bands and colors mostly due to ammonia in these upper layers (*see* Fig. 7.20). The feature prominent in the lower half of the image, not unexpectedly known as the Great Red Spot, is a storm system that has existed for hundreds of years (*see* Fig. 7.21). In the ever-changing face of Jupiter we can also now see, in addition to the Great Red Spot, Red Spot Junior and Baby Red Spot.

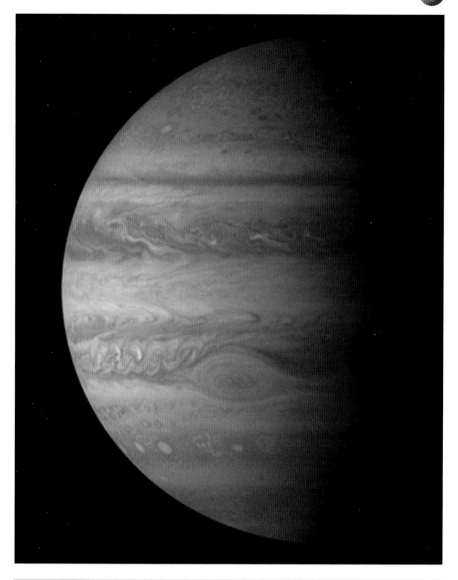

Fig. 7.20. True color mosaic of images from *Cassini* (Credit: NASA/JPL/Space Science Institute).

Hydrogen becomes metallic when, at huge pressures of more than a million atmospheres, the atomic electrons swish like a soup around the nuclei, allowing an electric current to pass.

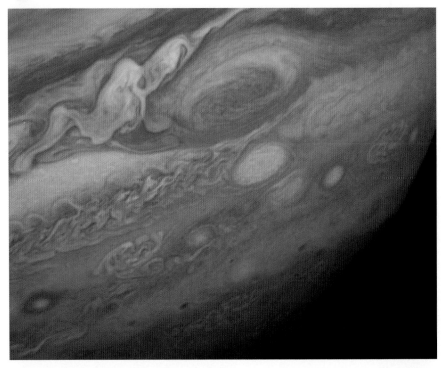

Fig. 7.21. Jupiter's Great Red Spot and complex cloud patterns (Credit: NASA).

Photographs from the *Voyager 1* spacecraft in 1979 surprised us all when they revealed large faint rings (*see* Fig. 7.22) extending out to 129,000 km from the center of the planet. A suggested cause is fine dust particles of less than 10 μ in diameter thrown up by meteoroid impacts on its moons. We will talk more about Jupiter's famous moons later.

Although scheduled to last 4 years to study Jupiter and Saturn, for more than 30 years since launch *Voyagers* 1 and 2 have continued to use their long-lived radio-isotope thermoelectric generators (300 W using alpha emitting plutonium-238, with a half life of 88 years) to transmit data to Earth about the solar wind, energetic particles, magnetic fields, and radio waves. Currently at the far reaches of the Solar System each craft, 12–15 billion km from the Sun, is still returning information, particularly concerning the interaction of our solar wind with the thin gas of space – the heliosheath.

Whereas the rocky planets have a surface, with much to be seen and photographed, the images of the surface of a gas ball are somewhat limited, though nonetheless worthy of frequent observation.

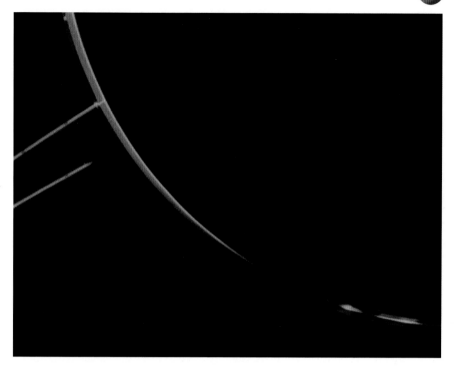

Fig. 7.22. Jupiter's faint ring system (Credit: NASA/JPL).

Saturn

The first sight of Saturn through a telescope (*see* Fig. 7.23) is a thrill-ing moment to most astronomers. Even small scopes reveal the planet's amazing rings, although they cannot be seen with the unaided eye. Gali-leo first spotted the rings, but the low resolution only allowed him to guess that they were large satellites. As telescopes became more and more powerful over the centuries more and more rings within rings were iden-tified. They are huge. The outermost ring is around 300,000 km across. In stark contrast their thickness is a mere 3,000 m at most and comprises billions of lumps of ice no more than 3 m across and many just the size of dust particles.

Saturn is not as large as Jupiter, but it is still a hydrogen/helium gas giant at around 120,000 km in diameter at the equator. With a density (0.70 g/cm^3) nearly a tenth that of Earth it would float if dropped into a large

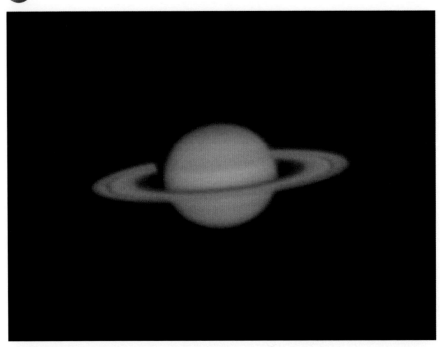

Fig. 7.23. Nice terrestrial view of Saturn (Credit: Mark Shelley, Orpington Astronomical Society).

enough puddle of water, the only planet that would do so. Saturn shows bands across its surface clouds similar to Jupiter but with less detail and structure. Because different-sized particles reflect, absorb, or scatter different wavelengths of light in a variety of ways detectors must select appropriately to determine the characteristics and constituents of the clouds. Hence false colors are needed to present a visible and meaningful picture such as the Hubble photograph, shown in Fig. 7.24, created through a combination of three images showing the planet in reflected infrared light. Different colors indicate the heights and composition of the clouds. A Hubble set of images (*see* Fig. 7.25) was created through the use of 30 selected filters on its Wide Field Planetary Camera 2 to illustrate the capability of multiwavelength detection and provide further data on the cloud composition.

It would have knocked Galileo's socks (sandals?) off to see what we see and know what we know now. He would have been staggered to be told that a spacecraft, *Cassini,* had actually been propelled there to take close-up pictures. One of the most stunning images of Saturn in natural

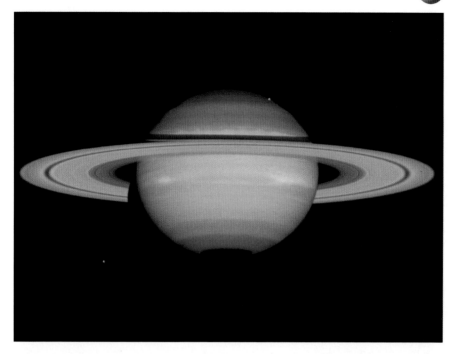

Fig. 7.24. An infrared view of Saturn (Credit: NASA/Erick Karkoschka).

color must be the one shown in Fig. 7.26. Red, green, and blue images were captured over a period of 2 h at 42 locations and tiled together for its creation.

A great example of imaginative technique combined with false colors is that of stellar occultation. Pointing one of the spacecraft's instruments through the rings at a star and converting the flickerings as the rings "passed" in front produced a density pattern exposing the huge number of rings and mini rings (*see* Fig. 7.27). It was thought that the rings were formed following a collision with one of Saturn's moons during the late heavy bombardment, 700 million years after Saturn was formed. Because the moon was orbiting at just the right distance, within the Roche Limit, when it shattered, the debris formed the rings instead of flying off in all directions. More recently it has been suggested that the rings are very much older than current estimates and will therefore have had a longer presence around the parent planet.

Here, now, are a few details concerning the *Cassini* mission. The Titan IV-B/Centaur launch vehicle carrying the *Cassini-Huygens* spacecraft was

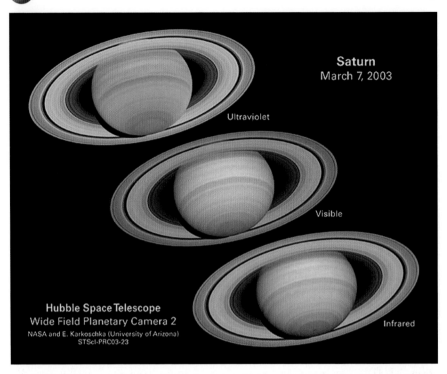

Fig. 7.25. The three color views of Saturn (Credit: NASA/Erick Karkoschka).

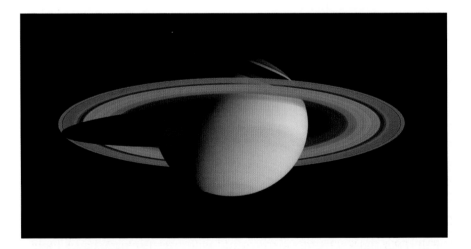

Fig. 7.26. Saturn natural color "Best Ever" picture (Credit: NASA/JPL/Space Science Institute).

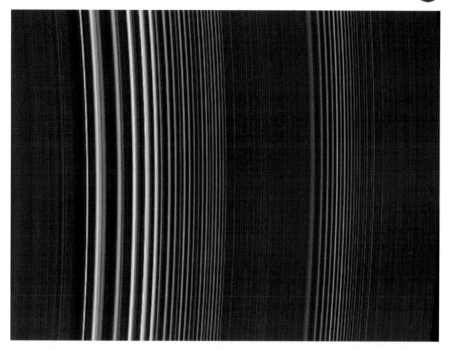

Fig. 7.27. False color image of Saturn's rings through stellar occultation (Credit: NASA/JPL/University of Colorado at Boulder).

launched from Earth on October 15, 1997, and arrived at Saturn in the summer of 2004. Scientists from 17 countries were involved in its construction. Gravity assists using the Sun, Venus, Earth, and Jupiter were necessary to propel the heavy spacecraft towards Saturn. Main engine bursts were activated to plant the craft in orbit around the planet on June 30, 2004, when the 12 scientific instruments, comprising imaging, spectrometry, and radar, could begin their data collection. The specific mission of ESA's Huygens craft was to float down to Titan and provide data on the atmosphere and surface – which it did most admirably.

The Roche Limit is the distance from the center of a planet within which any large natural body would be torn to pieces by tidal forces. The bits remaining inside the limit would disperse into rings and those outside would coalesce. The limit was first calculated by the French mathematician Édouard Albert Roche (1820–1883) in 1848.

Uranus

Uranus is different from other planets by virtue of its lying on its side, possibly due to an early cosmic collision, and its magnetic field lies tilted with respect to its rotational pole. Although not as stunning as Saturn, the first view of Uranus through a telescope is surprising! Being used to observing white or even reddish blobs it is surprising to see Uranus as a greenish or turquoise disc. Another large gas planet at around 50,000 km in diameter, Uranus owes its color to a layer of methane above the atmosphere of hydrogen and helium and has a liquid water, methane, and ammonia core. Sunlight is reflected from the cloud under-layer back through the methane that then absorbs red light to allow the blue-green color to be seen.

A surprise discovery was made in 1977. A star "near" Uranus appeared to blink several times, and it was realized that this was due to occultation of that star by previously unknown rings. Eleven rings have now been identified, though, once again, nowhere near the extensive content of those of Saturn. They are narrow, 5–100 km wide, and a mere 10 m or less thick.

An infrared composite image (*see* Fig. 7.28) obtained by the Keck ll Adaptive Optics system of the two hemispheres displays the vertical structure of atmospheric features. The highest clouds appear white, middle clouds bright green, and the lower clouds blue. The storms seen as small white spots are, in fact, huge and cover about 7 million square km. An artifact of the imaging process gives a reddish hue to the rings.

Neptune

Neptune has much in common with Uranus – its size, its composition, and its blue color. Once more, much of the blue derives from the absorption by methane, after reflection from the clouds beneath, of red light, although there may be another process that causes the extra blueness.

Just one spacecraft has visited Neptune, *Voyager 2* on August 25, 1989, but the Hubble space telescope has also provided pictures. Neptune is now the outermost planet, but Pluto's orbit is so eccentric that it crosses that of Neptune regularly, making the latter the outermost even when Pluto *was* a planet! Unexpectedly, although the axis of rotation is oddly parallel with the ecliptic, the polar regions pointing at the Sun are still colder than equatorial regions. Possibly partly due to the planet whizzing around in 17 h, the wind speeds are huge, up to 2,000 km/h and the fastest in the Solar System. In contrast to the Great Red Spot of Jupiter Neptune has a Great Dark Spot – actually "had" at the time of the *Voyager*

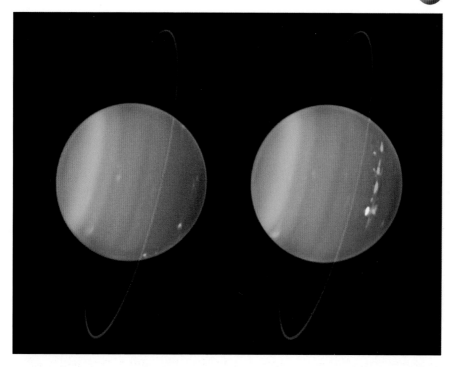

Fig. 7.28. Uranus and its rings (Credit: Lawrence Sromovsky, University of Wisconsin-Madison/W. M. Keck Observatory).

encounter, as it is no longer visible. A smaller White Spot zips around the planet in 16 h and has been given the nickname "The Scooter." Figure 7.29 shows the Dark Spot, center/left, the Scooter below, and the Dark Spot 2 with its bright center further below, lower right.

Much is unknown about the causes of many of the phenomena, and a mission to find some answers would be great.

Pluto

It is staggering that any pictures can be taken at all of this dwarf planet, because it is such a distant small rock, but these Hubble Space Telescope images, taken through a red and a blue filter, provide some information on the composition and origin of the satellites Charon, Nix, and Hydra (*see* Fig. 7.30).

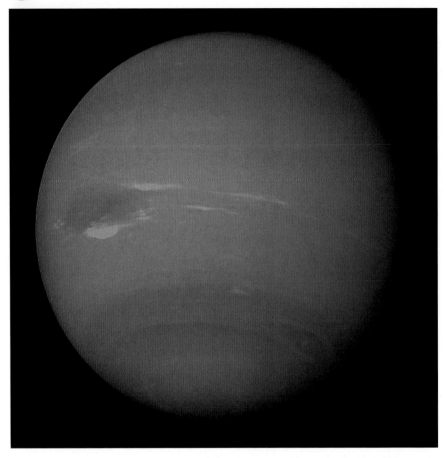

Fig. 7.29. Blue Neptune with its white and dark spots (Credit NASA/JPL).

The *New Horizons* mission began in January 2006, the objective being to reach the Pluto system in 2017. Already it has reached Jupiter and is scheduled to take a closer look there. It has explored details of the huge planet no mission has seen before, such as lightning at the poles and the life cycle of fresh ammonia clouds (*see* Fig. 7.31). The craft will not be returning to Earth but will continue beyond Pluto. Who knows what unexpected discoveries will be made? Perhaps more large rocks or ice balls from the Kuiper Belt to keep Quaoar, Sedna, and 2003 UB313 (now known as Eris) company? The Sloan Digital Sky Survey has detected a new object, 2006 SQ372, with a strange elongated orbit coming only as close as around the orbit of Neptune then out to 1,600 AU from the Sun.

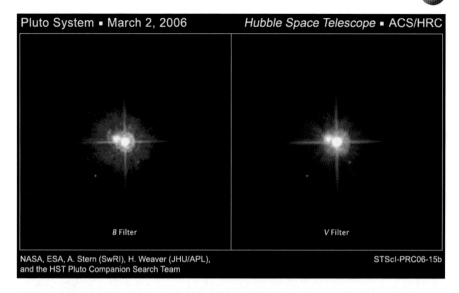

Pluto System ▪ March 2, 2006   *Hubble Space Telescope* ▪ ACS/HRC

B Filter   *V* Filter

NASA, ESA, A. Stern (SwRI), H. Weaver (JHU/APL), and the HST Pluto Companion Search Team   STScI-PRC06-15b

Fig. 7.30. Pluto's moons (Credit: NASA/ESA).

SQ372 may have come from an "inner" Oort Cloud. The Oort Cloud is believed to be a mass of icy, comet-like bodies surrounding the Solar System up to some 50,000 AU away from the Sun, or about a quarter of the distance to our nearest star, Proxima Centauri.

It has been proposed that some of the Oort objects might be nudged out of their environment toward Earth due to the gravitational influence of passing stars. Comets come under that description – dusty, icy lumps sweeping round the Sun and out again to return many years later unless terminated by a catastrophe or just worn out. The name *comet* comes from the Greek and means "long-haired," since, as the body comes closer to the Sun, the heat causes dust and gas to be released, forming a coma or tail.

Many comets visit the Solar System but are too dim to see. Because of ionization and the Sun's reflection, light is emitted, but only a few are bright enough to be observed by the unaided eye here on Earth. There are notable exceptions, such as the famous, long-period Halley's comet. Comet Hale–Bopp has been dubbed "The Great Comet of 1997" and was visible to the naked eye for 18 months at a maximum brightness to rival any star except Sirius. Hale–Bopp was particularly beautiful because it had a third tail, in addition to the usual gas and dust, composed of sodium (*see* Fig. 7.32).

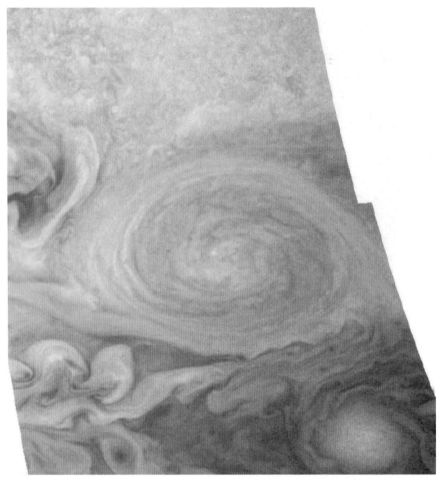

Fig. 7.31. Jupiter's clouds and Little White Spot (Credit: NASA/ John Hopkins University/HST).

As if to emphasize the unpredictability of arrivals by visitors from the outer Solar System, the periodic comet 17P/Holmes, first discovered by Edwin Holmes in 1892, returned in October/November 2007 (*see* Fig. 7.32a) with a sudden brightening thousands of times more intense than its usual level of brightness, making it visible to the naked eye. The circular shape, over a million kilometer across, was also unexpected and the outburst occurred near opposition, thus presenting a view along the axis of the tail.

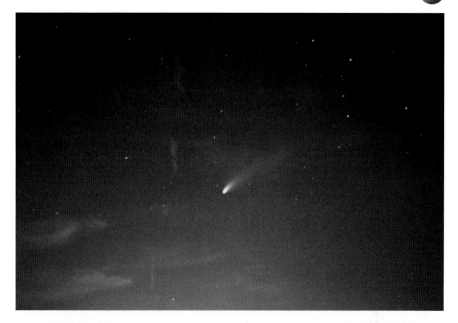

Fig. 7.32. Beautiful comet Hale–Bopp (Credit: Paul Whitmarsh, Orpington Astronomical Society). An easy image to capture of comet Holmes using an off-the-shelf digital camera and a modest telescope.

For observers in the Southern Hemisphere (also visible in the Northern Hemisphere) there was "The Great Comet of 2007," Comet McNaught, with a maximum magnitude estimated to be a staggering −6.0 and a 35° tail. A stunning image from an Aboriginal midden in Australia is shown in Fig. 7.33. Aboriginal shell middens are sites where the ancient Aborigines ate different kinds of shellfish.

Possibly even more exciting is the discovery of around 300 exoplanets, planets orbiting other stars, and some information concerning their composition and characteristics has been forthcoming from spectral analysis. A prolific discoverer of exoplanets is the European project WASP (Wide Angle Search for Planets) with two robotic observatories, one in the Canary Islands and one in South Africa. So far, many of these "planets" have shown themselves to be quite exotic in their size, density, and orbital characteristics that may reflect their stage of evolution.

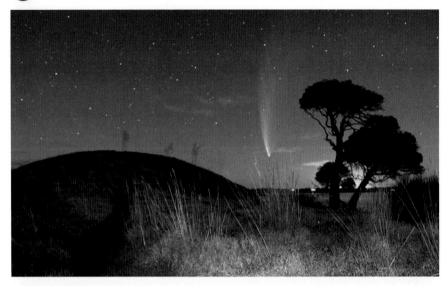

Fig. 7.33. Comet McNaught as viewed from an ancient Aboriginal shell midden (Credit: Peter Fuller, http://www.peterfuller.com.au).

Solar System Moons

Although the planets are the main bodies in our Solar System, their natural satellites, the moons, are not just fillers. They are a fascinating treasure of mystery and diversity in composition, origin, and color. New moons are being discovered at a tremendous rate, as the technology of telescopes and detectors becomes ever more sophisticated. The question then is often addressed as to how large a lump of cosmic debris has to be, to be classed as a moon? A few meters? A few kilometers? Tens or hundreds of kilometers? The topic of colors is especially appropriate for the larger satellites, where more detailed images are possible for presentation of pleasing true or false color pictures. Just a selection of the myriad satellites is presented here.

Neptune has 11 known satellites, the largest being Triton, at around 2,705 km in diameter. Of the major moons it is the only one that orbits in a direction opposite to that of its parent planet, suggesting it might be a captured comet. At −235°C it is the coldest known temperature for a body in the Solar System, even shooting crystals of nitrogen ice from its remaining volcanic activity (Table 7.1).

Figure 7.34, a false color image taken by *Voyager 2* through ultraviolet, violet, and green filters, indicates the probable volcanic activity and methane and nitrogen in its thin atmosphere. Evidence has also been

Table 7.1. Boiling points of some commonly encountered gases in the solar system

Ammonia	240K	−33°C
Argon	87K	−186°C

Carbon dioxide cannot be a liquid at atmospheric pressure. It is solid below −78.51°C and turns directly into a gas above that temperature, in a process known as sublimation.

Carbon monoxide	81K	−192°C
Ethane	184K	−89°C
Helium	4K	−269°C
Hydrogen	20K	−253°C
Methane	112K	−161°C
Nitrogen	77K	−196°C
Oxygen	90K	−183°C

found that volcanoes on Triton once gushed a slushy mixture of water and ammonia. Figure 7.35 shows a natural color image, although the colors are somewhat exaggerated, indicating Triton is mostly white with a bit of pink in places. *Voyager's* natural color image of Triton, Fig. 7.35, is a composite of pictures taken through violet, green, and clear filters when just 210,000 km away from the wickedly cold satellite. The pockmarked moon presents a rough, pitted surface in one area and a smoother one in another. The pale pinkish hue indicates a thin coating of some material, possibly a frost of methane, carbon monoxide, or nitrogen.

Temperature measures the energy of moving atoms or molecules. The faster the atoms or molecules move the higher the temperature. As they slow down the temperature decreases, so when they stop the temperature is at its absolute zero. That temperature is called 0 Kelvin and is equivalent to around −273°C. In the outer reaches of space where it is dark and with the smallest particle or gaseous content the temperature still isn't zero. It is 2.7 because of the 3° microwave background that is everywhere – always.

Of Uranus's 30 or so satellites there are five major ones – Miranda, Ariel, Umbriel, Titania, and Oberon – with diameters ranging from 500 to 1,500 km. *Voyager 2* took its best color picture of Miranda on January 24, 1986, through its narrow angle camera's green, violet, and ultraviolet filters. Figure 7.36 shows clearly the marked geological variation of different sectors with their ridges, valleys, and impact craters.

Voyager's best color picture of Ariel, Fig. 7.37, was taken from a distance of 170,000 km and was captured through the narrow angle camera's green,

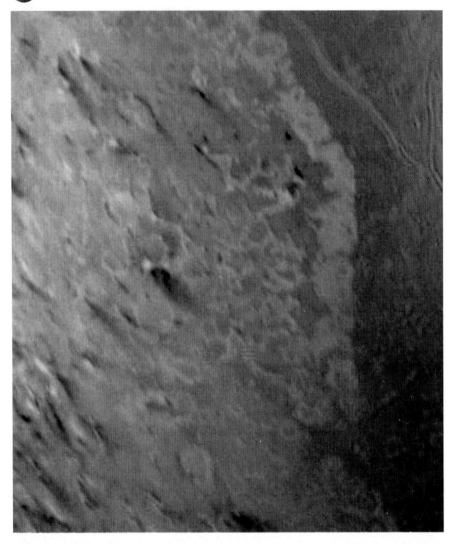

Fig. 7.34. *Voyager 2's* false color image of Triton (Credit: NASA/JPL).

blue, and violet filters. The surface is intensely pitted, with bright-rimmed craters showing at the edge of the moon and large faults and scarps. Oberon (*see* Fig. 7.38) clearly shows huge impact craters, especially the one in the center with an extensive ray system.

Saturn is parent to about 60 satellites or moons, of which eight are the major bodies, mostly with diameters in excess of 500 km. Of course huge Titan, at over 5,000 km in diameter and one of the largest moons in the

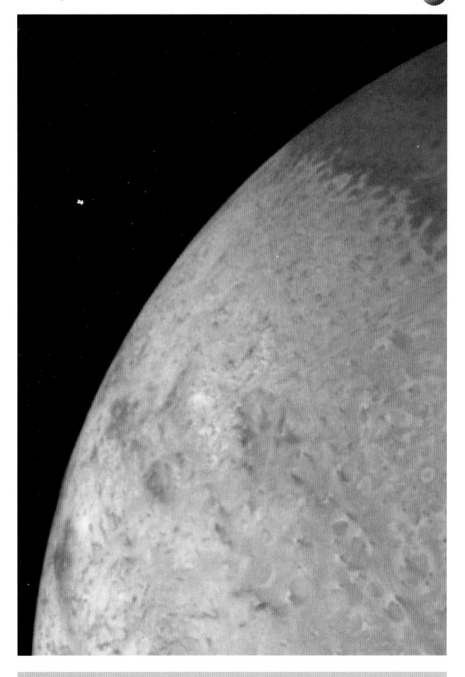

Fig. 7.35. *Voyager 2's* natural color image of Triton (Credit: NASA/JPL).

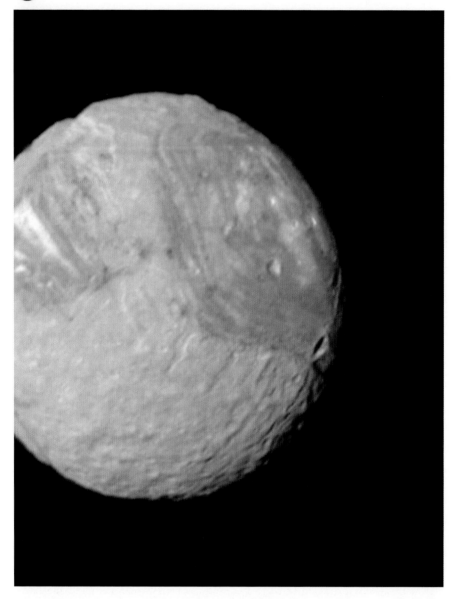

Fig. 7.36. *Voyager's* image of Miranda (Credit: NASA/JPL).

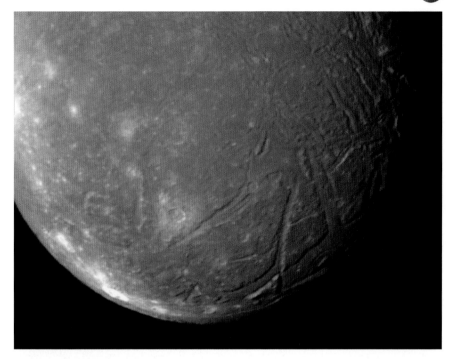

Fig. 7.37. *Voyager 2's* image of Ariel (Credit: NASA/JPL).

Solar System, has been studied intensively during the *Cassini/Huygens* exploration – a wonderful example of what people can achieve through collaboration and dedication.

The color filters red, green, and blue were employed in photographing this stunning Titan backdrop to a portion of the Saturnian rings, with battered Epimethius thrown in; this image is "completely artificial," although approximating to what a human might see (*see* Fig. 7.39). Completely unknown prior to the mission was what lay beneath the mysterious clouds of Titan. The false color image, clearly false as the acquiring instrument was *Cassini's* radar, staggered everyone when surface features similar to Earth such as lakes, islands, seas, and inlets, were discovered the big difference being their composition. Most likely what is water here on Earth is liquid methane and ethane there (*see* Fig. 7.40).

Exploration of the Saturn-facing side of Tethys, just over 1,000 km in diameter, benefits from this pair of true and extreme false color images, possibly indicating the variation of composition or particle size of the terrain (*see* Fig. 7.41). Ultraviolet, green, and infrared images were combined and overlaid onto a clear-filter image that preserves the relative brightness

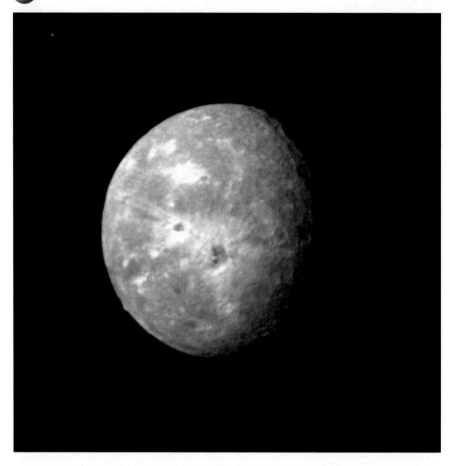

Fig. 7.38. Oberon from *Voyager* 2 in 1986 (Credit: NASA/JPL).

across the satellite. The deeply grooved surface at the top of the image, Ithaca Chasma, contains the crater Telemachus with its central peak.

Galileo made Jupiter's moons famous. As he studied the sky with his new-found telescope from 1609 he spotted four of Jupiter's largest satellites that are commonly referred to as the Galilean moons. Ganymede, Callisto, Europa, and Io are among the largest in the Solar System, and Ganymede is THE largest. Every moon within the Solar System is fascinating and unique with its own story to tell, but the focus here is on color, both natural and how false color can bring out features and composition.

Voyager 1 captured this wonderful color picture of Ganymede (*see* Fig. 7.42), showing impact craters and groove-like structures. Craters with bright rays often stretching 500 km or more have a definite bluish hue that appear white against the reddish background. The image in Fig. 7.43

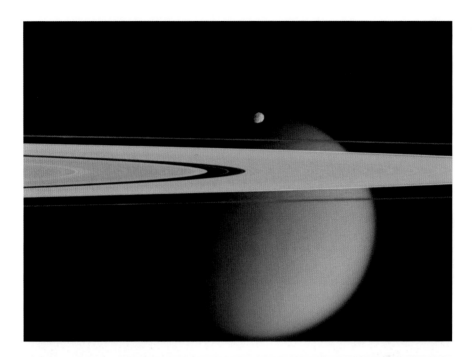

Fig. 7.39. Titan through Saturn's rings (Credit: NASA/JPL/ Space Science Institute).

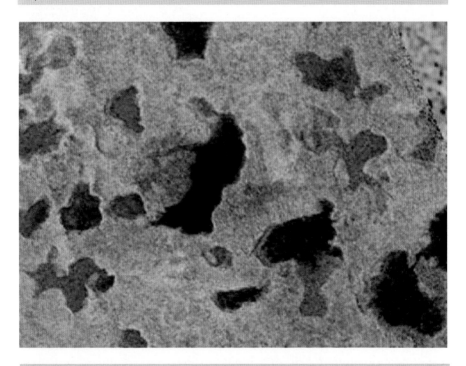

Fig. 7.40. Radar image of Titan's lakes and seas (Credit: NASA/JPL).

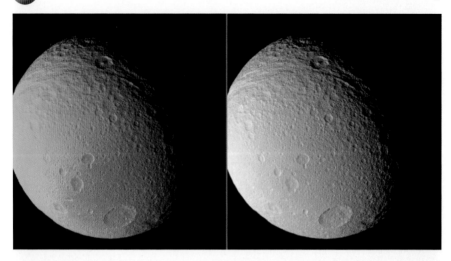

Fig. 7.41. True (*right*) and false color images of Tethys (Credit: NASA/JPL/Space Science Institute).

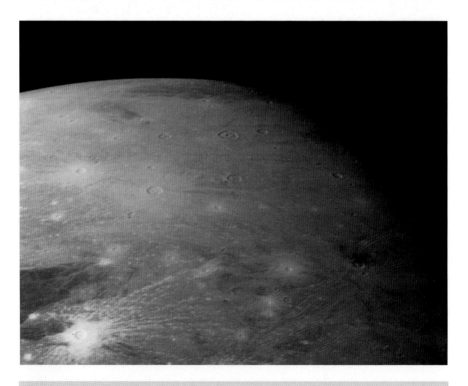

Fig. 7.42. *Voyager's* image of a region in the northern hemisphere of Ganymede (Credit: NASA/JPL).

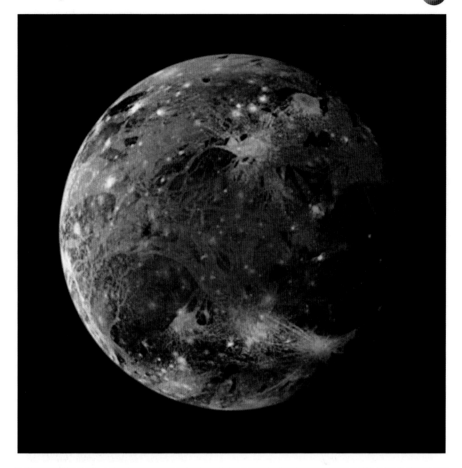

Fig. 7.43. Spooky Ganymede (Credit: NASA/JPL/DLR).

has an almost spooky appearance and is the result of color enhancement of a photograph taken by the *Galileo* orbiter in 1998. Frosty polar caps, bright grooved terrain, and older, darker furrowed areas are revealed by the enhancement.

Jupiter's Callisto is the third largest moon in the Solar System after Ganymede and Titan. The image in Fig. 7.44 is the only complete color image of Callisto taken from the *Galileo* orbiter showing its uniform scattering of craters but nonuniform distribution of brightness and colors, thought to be due to the presence or lack of ice. The value of image colors is well expressed in the double image shown in Fig. 7.45, comprising an approximate natural color on the left and a false color image on the right to enhance subtle color variations. The huge impact feature Valhalla, just above center, is around

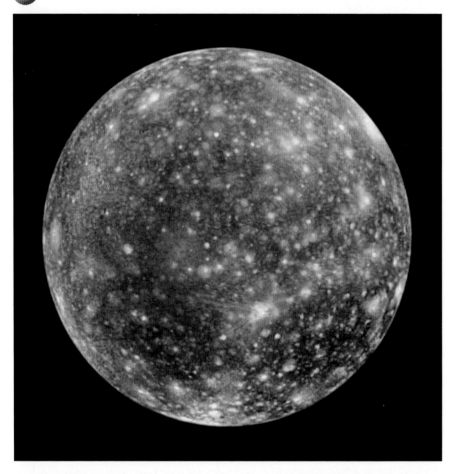

Fig. 7.44. Complete global color image of Callisto (Credit: NASA/JPL/DLR).

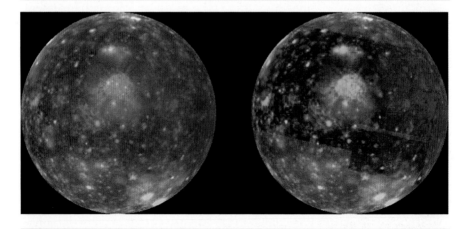

Fig. 7.45. Callisto in true and false color (Credit: NASA/JPL/ University of Arizona).

600 km across, the brightness possibly being due to uncovering of subsurface material following impact. NASA's own words describe comparison of the two images and the color selection:

> The false color in the right image shows new information, including ejecta from relatively recent craters, which are often not apparent in the natural color image. The color also reveals a gradual variation across the moon's hemisphere, perhaps due to implantation of materials onto the surface from space. The false color is created from ratios of infrared/violet and its inverse (violet/infrared), which are then combined so the infrared/violet, green, and violet/infrared are assigned to red, green, and blue in a composite product.

There is much still unknown about the lenticulae on the surface of Europa, the smallest of the Galilean moons, looking in this enhanced color image from the *Galileo* orbiter more like a medical disaster (*see* Fig. 7.46). (Lenticula is the Latin word for freckles as well as for a small lens.) The dents and bumps in the picture are about 10 km across. The presence of much surface and subsurface water has led to speculation concerning the existence of primitive life there.

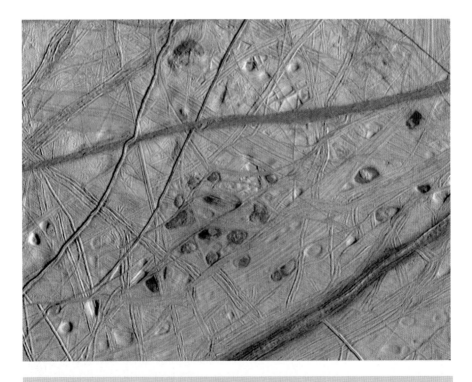

Fig. 7.46. The cracked surface of Europa (Credit: NASA/JPL/ University of Arizona/University of Colorado).

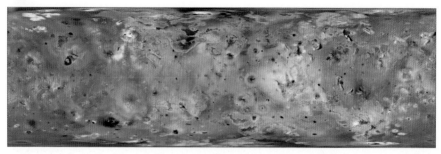

Fig. 7.47. Io, the most volcanically active body in the Solar System (Credit: NASA/JPL/USGS).

Less a picture of a medical disaster and more of an extreme omelette is the surface of Io, which is constantly subjected to gravitational pressures from its parent Jupiter. Massive volcanic eruptions of silicate lava and sulfur contribute to the color of the constantly changing surface. The image shown in Fig. 7.47 in almost natural color is a combination of the best images from the *Voyager 1* and *Galileo* missions. Color information from the *Galileo* violet, green, and near infrared images was superimposed upon a monochrome mosaic of the highest resolution images from *Voyager* and *Galileo*.

Mars's two moons are tiny. Phobos and Deimos are just 21 and 12 km in diameter, respectively. Although initially difficult to image, they were studied mainly through spectroscopy and were thought to be a rather dull gray, leading to a hypothesis that they might be carbonaceous chondrite material originating from the outer Asteroid Belt. More recent photographs have shown them to be somewhat reddish, as illustrated by false color pictures (*see* Fig. 7.48), and more closely resemble ultraprimitive bodies in the outer Solar System. Both images were captured by the Compact Reconnaissance Imaging Spectrometer for Mars (CRISM) in 544 colors, covering 0.36–3.92 μm.

A chondrite is a specific variety of stony meteorite that has not been modified by melting or differentiation of the parent body. Chondrites were created when various types of dust and small grains that were present in the early Solar System accreted to form primitive asteroids.

Since neither Mercury nor Venus have their own natural satellites, our Moon is the final beauty to consider. And what a beauty! Its proximity to Earth, its regular appearance, huge size, and multitude of unique and fascinating surface features make exciting images accessible to all even with an off-the-shelf digital camera or binoculars. Obsessions with the Moon

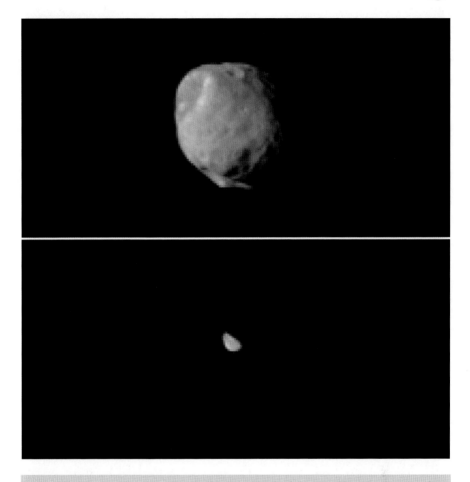

Fig. 7.48. Mars's two small moons, Phobos and Deimos (Credit: NASA/JPL/John Hopkins University Applied Physics Laboratory).

by astrophotographers is commonplace, so we will offer no apologies for the number of images included here. A typical example of an awe-inspiring feature, easily obtainable using a modest telescope and compact digital camera, is shown in Fig. 7.49. Such images can become quite personal (no one could have seen this before, surely!!!) and responsible for the beginning of a lifelong pursuit of better and better images. So much has been published by way of facts and photographs of the Moon that an attempt will be made to restrict descriptions to aspects of color.

The color of Moon pictures taken from Earth is inevitably influenced by our refracting and particle-laden atmosphere (*see* Fig. 7.50). Seen from

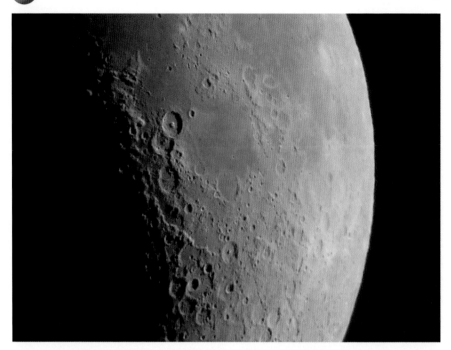

Fig. 7.49. Beautiful craters and a scarp surround Mare Nectaris.

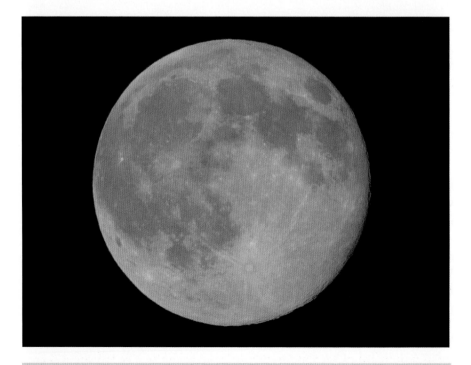

Fig. 7.50. Full Moon with the prominent Tycho "recent" impact crater.

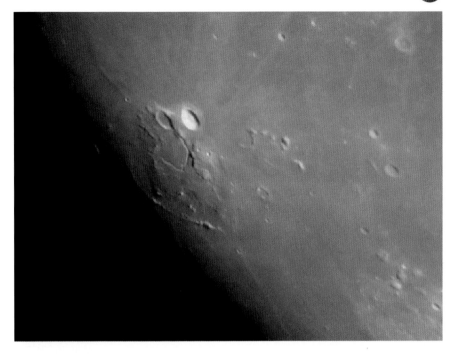

Fig. 7.51. Color and detail of the Aristarchus area.

space, and without enhancement or processing, natural color images of the Moon depict, essentially, a landscape devoid of color, mostly shades of gray with one or two areas such as northwest of Aristarchus (Wood's Spot or Aristarchus Plateau) showing a discernable reddish hue, first noticed in 1647 by Hevelius. Figures 7.51–7.53 show a lunar panorama containing the Aristarchus Plateau, the reddening accentuated through addition of a little computer photographic enhancement and an enlargement to showcase the fascinating detail of the Vallis Schröteri within the Aristarchus area. (The trick with these pictures is in knowing how to reveal the colors, which can be achieved even from amateur digital camera and webcam photographs. Once revealed they present a wealth of information about the geology and chemistry of the surface.)

The key word is saturation, not the usual perception of the meaning, as in excessive loading with water, but the vividness or purity of a color. When an object reflects light and absorbs every color except, say, green, then that green is said to be highly saturated. If the object absorbs some of that green along with everything else then the green is less saturated. The more blacks and grays that appear in the photo the less saturated it is. However, the more saturated the less detail, so it is important to balance the requirement of resolution with bringing out the colors. Cameras can

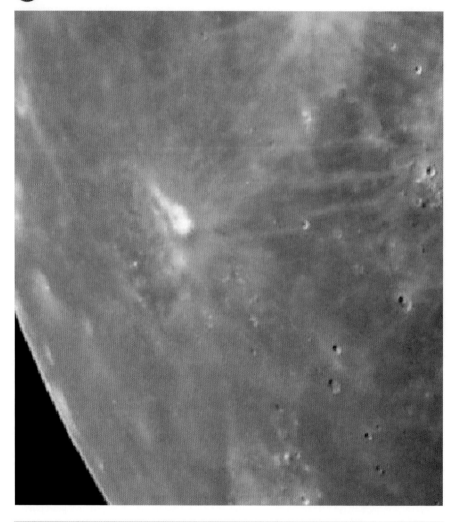

Fig. 7.52. Color and detail of the Aristarchus area.

be preset for saturation, or, more commonly, the images are processed later using image manipulation software. More details of saturation and instructions on how to capture the colors of the Moon are given by Filipe Alves below. The two wonderful images in Fig. 7.54 and the glorious *Galileo* image in Fig. 7.55 show what can be achieved.

Areas appearing red generally correspond to lunar highlands, while blue to orange shades indicate the ancient volcanic lava flow of a mare, or lunar sea. Bluer mare areas contain more titanium than do the orange regions. Mare Tranquillitatis, seen as a deep blue patch

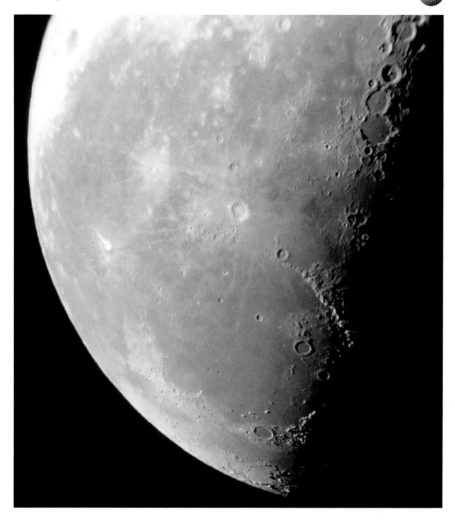

Fig. 7.53. Color and detail of the Aristarchus area.

on the right, is richer in titanium than Mare Serenitatis, a slightly smaller circular area immediately adjacent to the upper left of Mare Tranquillitatis. Blue and orange areas covering much of the left side of the Moon in this view represent many separate lava flows in Oceanus Procellarum. The small purple areas found near the center are pyroclastic deposits formed by explosive volcanic eruptions.

ESA's mission SMART-1 (Small Missions for Advanced Research in Technology) had many purposes, including the testing of an ion drive propulsion system. An important instrument on board was the D-CIXS,

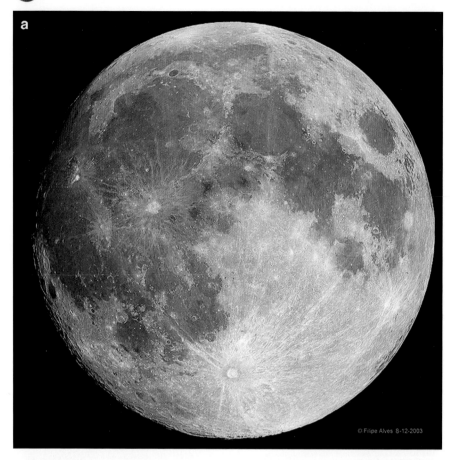

© Filipe Alves 8-12-2003

Fig. 7.54. Application of saturation (Credit: Filipe Alves).

Demonstration Compact Imaging X-ray Spectrometer, designed to observe how the Moon fluoresces in X-rays when the Sun shines on it. Not only did they quickly see the elements aluminum, silicon, and iron but, crucially, calcium. Much information about the geology and hence the origin of the Moon has been gained through consideration of such data in combination with analysis of rocks brought back by the lunar landing missions. Images from SMART are to be found at the ESA site referred to in the Appendix of this book.

Luckily, just as the scientists took their first look at the Moon using D-CIXS, a solar flare occurred, making it easier for D-CIXS to collect its first set of data on Friday June 10, 2005.

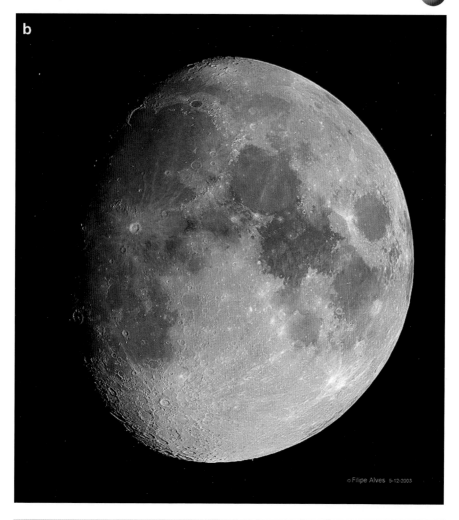

Fig. 7.54. (Continued)

Images captured by amateur astronomers are often black and white (or gray scale) for optimum resolution, but the atmospherically affected color photograph can look very picturesque and emphasize the variation in landscape features. The image shown in Fig. 7.56 is a good example and was taken through a 200 mm Celestron reflector with an off-the-shelf compact digital camera fixed to a position above the eyepiece. The Apennine Mountains (Montes Apenninus) near the terminator around first quarter beautifully reflect the "morning" sunlight to pick out the

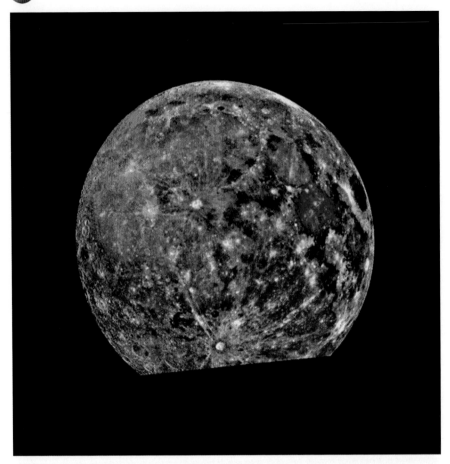

Fig. 7.55. Our Moon in color (Credit: NASA/JPL).

intricate furrows and boulders among the majestic peaks, preserved since ancient times by the airless and almost waterless moonscape. The identifying "tails" of craters of Archimedes and the teardrop Eratosthenes are exhibited well at these conditions of illumination. The gradation of light and dark reddish-brown hues brings out the features due to volcanic ash, rays from crater Manilius, and the ridge and crater shadows. References to more detailed instructions on how to take such photographs and identify the geography are given in the Appendix of this book.

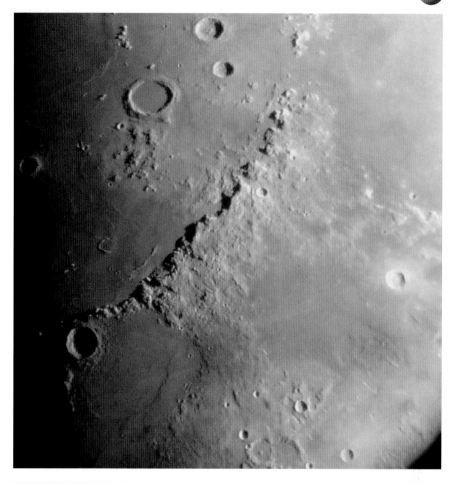

Fig. 7.56. Morning sunlight on the Apennine mountains.

Weather permitting, photographing a lunar eclipse is a most satisfying experience. It can be almost black/invisible, but March 2007 conditions allowed perfect viewing to capture this image (*see* Fig. 7.57). The amount of redness (on the Danjon scale) as a result of the extent of scattering of blue light in the Earth's atmosphere is always exciting to anticipate. More pictures of the Moon and its many "moods" are included in a later chapter.

The Danjon scale is a 5-point scale ranging from 0 at its lowest, i.e., Moon very dark or invisible, to 4 – extremely bright.

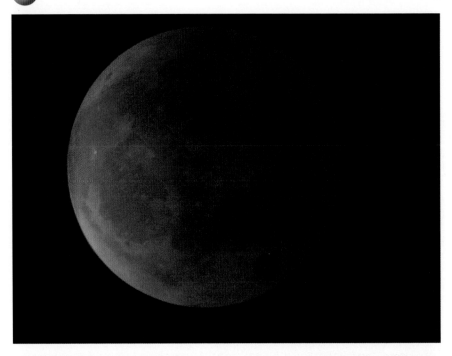

Fig. 7.57. Lunar eclipse, 2007.

CHAPTER EIGHT

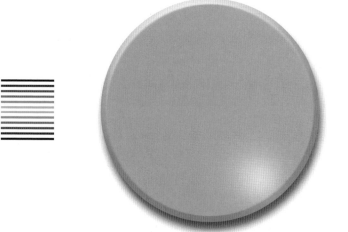

Colors in Earth's Atmosphere

To capture records of unusual or unexpected events, "be prepared" is the most important watchword, second only to opportunity, which is usually out of your control. How many times do we see a colorful event in the sky or on the ground and say "I wish I had my camera with me!" It has been said that, in temperate climes, we are only privy to about ten rainbows per year, but, if you spend lots of time gazing out of window you might notice many more rainbows and part rainbows that suddenly appear, then vanish after a few seconds, and colorful sunbeams that race across the sky on a windy day; all very easily missed and excluded from a count of phenomena. The message is clear. Like a Wild West gunslinger your camera must be available to be drawn out of its holster at a moment's notice.

One contact in America said that she never goes anywhere without her camera, even to take out the garbage! Not only that, but without warning of any particular events, the menu must be set to "average" or "normal" settings. I have a wonderful green movie of a cement mixer (don't ask!) simply because the last camera use involved a test of vivid colors.

However, there are times when opportunity gives a little warning; the thunder and lightning storm that rumbles on for a while and the overhead red shimmer and cloud positions heralding the possibility of a photogenic sunset. Contrast this with a Sun pillar or a Moon halo, often past their best within a few seconds. It also seems that the rarer the event the more stunning we say the photograph appears.

T. Buick, *The Rainbow Sky*, Patrick Moore's Practical Astronomy Series, DOI 10.1007/978-1-4419-1053-0_8, © Springer Science + Business Media, LLC 2010

Knowing how to get the best out of your camera is vital. A beautiful and ephemeral flower that was known to last for only 1 day was just asking for a photo to be taken on a bright and sunny day. "Snap" or auto settings were allowed to prevail as the camera was "clicked." The picture in Fig. 8.1 shows a nice home photograph from a hand-held camera. However, applying a few simple settings to improve the sharpness revealed a startling difference as presented in the enlarged images in Fig. 8.2. You just might wish to frame an enlarged picture sometime and be pleased that a little extra care was taken.

The simple settings applied in this example were: use a tripod, select timed shutter release so that there is no "hands-on" movement, select close-up or macro, and ensure the camera is as parallel as possible to the plane of the flower or area of interest.

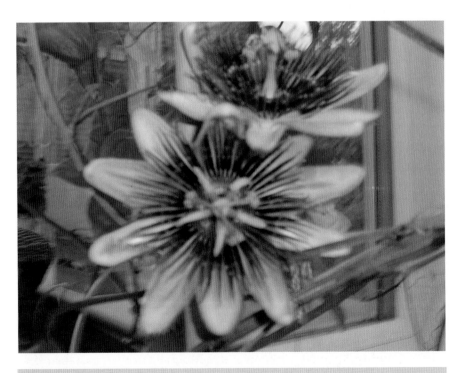

Fig. 8.1. A "snap" photo of a colorful flower (credit: Wendy Poole).

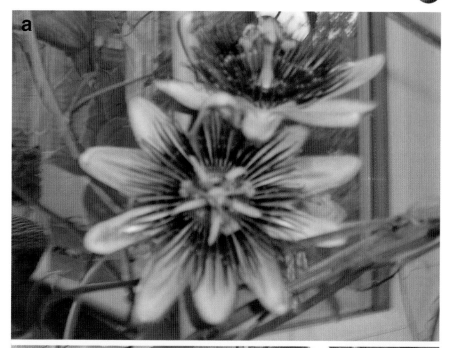

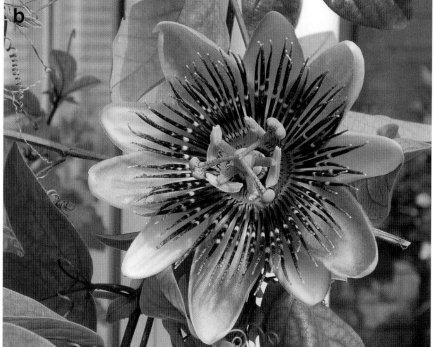

Fig. 8.2. The huge difference in resolution between auto settings and a few simple manual ones (credit: Wendy Poole).

Most of the colorful atmospheric sights commonly listed for amateurs to capture are a result of reflection, refraction, or a combination of the two effects in direct or indirect sunlight – or at least the narrow portion of the electromagnetic spectrum that activates electrochemical responses within the rods and cones of the human retina. The variables are the backdrop, the source of illumination, and the medium and its structure that all have an effect on component wavelengths.

"You mean the weather!" Er! Yes!

Although water in its various states – ice, rain, and clouds – is predominantly responsible for many of the spectacular colors, airborne dust can also contribute. Aerial pollution is a global problem, as an increasing number of countries become industrialized and belch out gasses and particulates of all sizes. There are also accidents and natural polluting events, such as forest fires and volcanoes, that can be responsible for adding extra hues to the night or day sky. Following the eruption of Mount St. Helens (see Fig. 8.3), in May 1980 an estimated 540 million tons of ash fell over an area of more than 60,000 km².

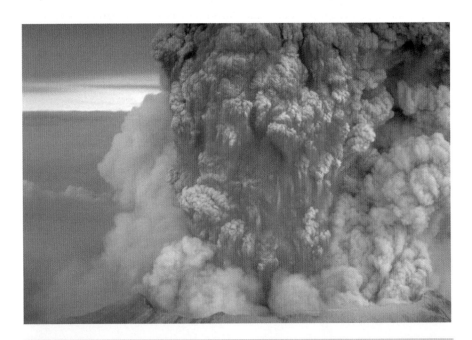

Fig. 8.3. Eruption of Mount St. Helens (credit: US Geological Survey).

The arrival of digital cameras has generated an explosion of the number of amateur photographers, and there are huge numbers of websites showing just what can be achieved with dedication and often a little help from computer image-manipulation applications. The ability to take any number of photographs at negligible cost and with instant feedback allows rapidly changing colors in the sky to be recorded when at their best.

Capturing the best photograph benefits from knowing what to do combined with the experience of successes and learning (failures). As noted earlier, if time and accessibility allow, then a tripod is essential to obtain the sharpest pictures. There is often enough movement going on at the event itself without adding the inevitable shakes of hand-held equipment. However, rather than miss an event, any means and any camera will do to prove you are witnessing an unusual occurrence. Even mobile phone cameras can produce acceptable records, especially as the specifications for combined phones and cameras are increasing at a fast pace. For some of the more ambitious targets a telescope plus digital camera can, with a little guidance, be very simple to operate for satisfying results. Although point-and-shoot techniques are frequently successful, a full knowledge of and familiarization with a camera's facilities will often ensure that the best possible photograph has been obtained. Not all compact cameras will have zoom, spot metering, long time exposure, remote shutter operation, etc., but what a shame if they were necessary, available in your camera, and you did not know!

Bows and Dogs

Of course, rainbows immediately spring to mind when thinking about colorful events, and beautiful they are. But there are icebows, fogbows, dewbows, part-bows, spraybows, and more such as circumzenithal arcs, circumhorizontal arcs, parhelia (Sun dogs), and parselene (Moon dogs). Some are formed from the interaction of light with water and some with ice, others are reflected or refracted onto or from clouds at different altitudes, and some involve particulates and substances such as sulfuric or nitric acid. Haloes at middle latitudes are relatively common, but rarer forms are seen in the Antarctic, due to very fine ice crystals known as "diamond dust" suspended in the air. Everyone must have seen a few "bows" at least in their time, and they occur in a great variety of situations.

Among the most common are the stunning arcs across the sky (see Fig. 8.4). Although the rainbow colors are a continuous spectrum (see Fig. 8.5), it has been convenient to quote them as seven discrete bands appearing in the order red, orange, yellow, green, blue, indigo, and violet already described with regard to the electromagnetic spectrum. Sir Isaac Newton added the unexpected "indigo" to ensure seven colors to link them with the seven notes in the musical diatonic scale and the seven known bodies in the Solar System – the Sun, Mercury, Venus, Earth, Mars, Jupiter, and Saturn.

To understand the generation of a rainbow a blanket of rain can be considered as an accumulation of what is happening in a single round drop of water (see Fig. 8.6). As light enters through the surface it transfers from air to water. A change in the media through which the light travels causes refraction or dispersion, a separation of the white light into its constituent colors, because different wavelengths are refracted to different degrees. The separated colors then hit the back of the drop and are reflected out of the front of the drop at an angle of around 40°. As the light emerges it is refracted again, reinforcing production of the spectrum, or rainbow. With the Sun behind and the rain in front of the observer, and at the right position for the 40° between Sun and viewer, the colored bow will appear. That is the simple explanation.

In fact, there are many light paths entering and leaving the drop, but there is a particularly intense beam exiting at the 40° angle that leads to a rainbow. The diagram shows a single internal reflection. A double internal reflection results in a weaker bow, with the colors reversed, as can be seen in the photographs, i.e., a primary rainbow has violet to the inside of the arc and a secondary to the outside. The rainbow is, of course, a complete circle, but only half is visible unless observed from high up, such as from a mountain or an airplane.

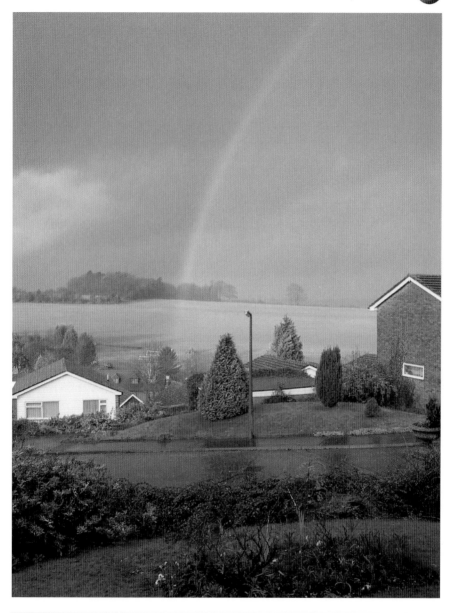

Fig. 8.4. A towering colored arc looms over a nearby residence.

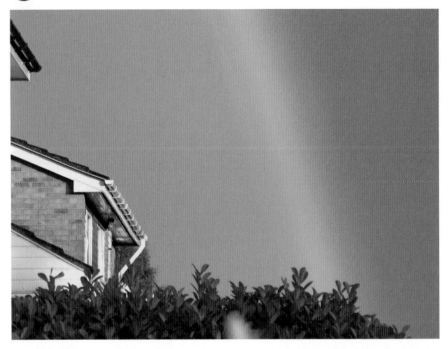

Fig. 8.5. Zooming in on the rainbow reveals the continuous nature of the spectrum, not discrete color sections.

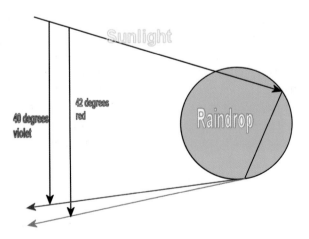

Fig. 8.6. Light path for a single reflection inside a raindrop leading to dispersion.

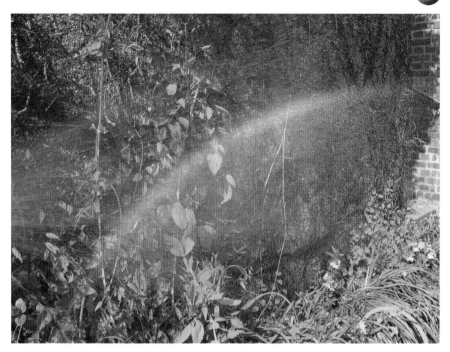

Fig. 8.7. A water spray on a sunny day demonstrates the formation of a rainbow (credit: Ellen Thompson).

It is easy to illustrate the formation of a rainbow for yourself by spraying water, on a sunny day, against a darkish background. The importance of the position of the observer and the angles involved becomes clear (see Fig. 8.7). Quite exciting is the surprise of a rainbow touching the ground and racing toward you across the fields, although, of course, it will never reach you, as a rainbow is an optical effect that depends on the position of the viewer relative to the Sun and raindrops (see Fig. 8.8).

Not quite so common is a really bright rainbow that displays a secondary one outside the primary at a slightly different angle of around 50° as a result of a second reflection inside the drop that also causes inversion of the order of colors. Since no reflected light emerges between the angles of 40° and 50° the space between the primary and secondary rainbows appears a little darker. This region is called the Alexander band (see Fig. 8.9). Higher order bows are rare and not so easy to spot.

Sun dogs or Mock Suns (proper name parhelia) may look like small strips of a rainbow, but they are very different in origin. For a start they

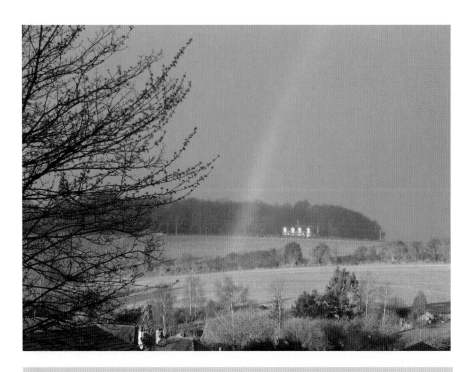

Fig. 8.8. The exciting spectacle of a rainbow racing towards you.

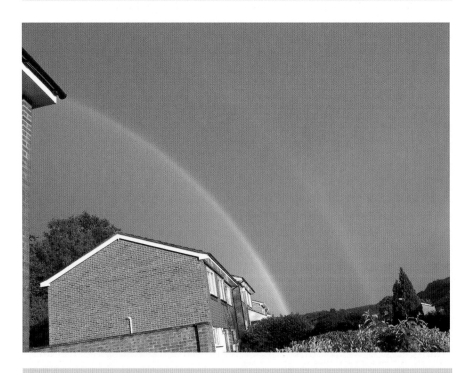

Fig. 8.9. The dark Alexander band on the red side of the rainbow can be seen contrasting well with the brighter violet side.

can be seen in the same direction as the Sun, not opposite. They are quite common and, if cloud formation is appropriate, appear in pairs approximately 22° either side of the Sun and at the same altitude above the horizon. They are often seen sitting on a solar halo with the red part of its spectrum towards the Sun on the inner edge while the blue part, often seen with a streaming white tail, is on the outer edge.

From around A.D. 1,000 many good theories explaining the production of rainbows were expounded, but René Descartes, in 1637, was the first to fully describe how they are formed. However, the dark band was named after the philosopher Alexander of Aphrodisias who lived in the second and third century A.D. and first described the darkness.

Disappearing when the Sun's altitude is above 60°, Sun dogs are associated with cirrostratus clouds that are composed of hexagonal ice crystals that reflect and refract sunlight (see Fig. 8.10). Compact cameras, with automatic focus built in, sometimes have difficulty coping with moving or fuzzy clouds. Therefore, it is helpful to include a suitable object in the frame, such as a tree (not too close but not out of the depth-of-field focus) on which to focus that may also enhance the presentation of the picture.

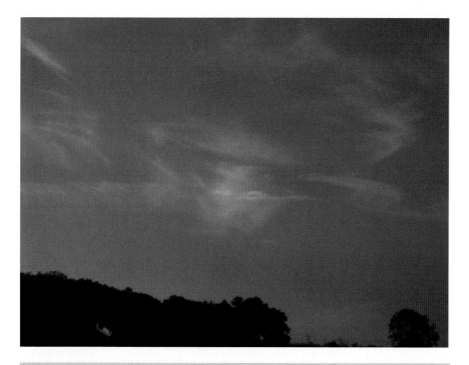

Fig. 8.10. Just enough cloud to allow appearance of a Sun dog.

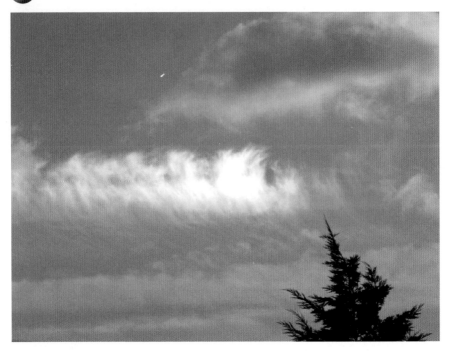

Fig. 8.11. A nice image of a Sun dog.

If available, set to "spot metering," otherwise a somewhat overexposed image might happen that could wash out color detail (see Figs. 8.11 and 8.12). Once a photograph is obtained it is very pleasing to zoom in for a closer look at the wonderful colors.

The photographs in Figs. 8.13–8.15 captured the long white stream in addition to some vivid hues. With the right cloud formation a twin Sun dog might appear (see Fig. 8.16). Figure 8.17, lower right, is a Sun dog captured over the hills of New Zealand.

Closely related to parhelia are haloes that are also formed from refraction and reflection of light through ice crystals in cirrus clouds. A Moon halo is not uncommon, but a complete circle is outside the field of view of many cameras, and separate frames need to be stitched together. Figure 8.18 shows the incomplete circle. Figure 8.19 shows the full circle consisting of several frames stitched together. Photographs rarely do justice to such haloes. The real thing is awe inspiring, with boundaries easily picked out by the eye. A parselene, or Moon dog or Mock Moon, only occurs when the Moon is bright, around full Moon, and since parselenes are much fainter than Sun dogs they usually appear colorless. Moon dogs are often the culprits of reports of UFO sitings!

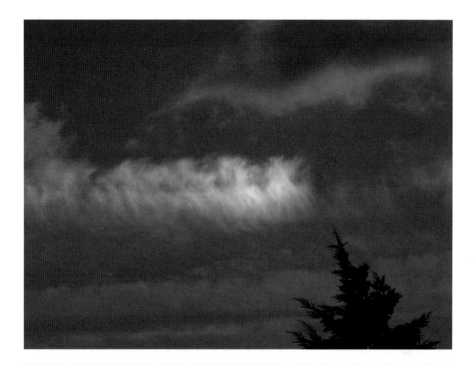

Fig. 8.12. Colors are brought out a little better with the application of spot metering.

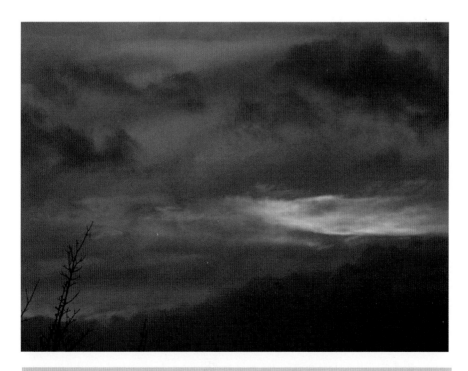

Fig. 8.13. An almost infinite variety of Sun dogs can be observed and photographed.

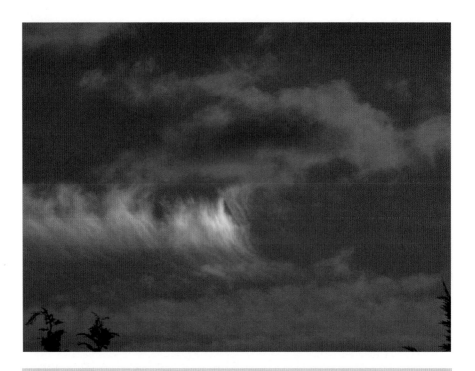

Fig. 8.14. An almost infinite variety of Sun dogs can be observed and photographed.

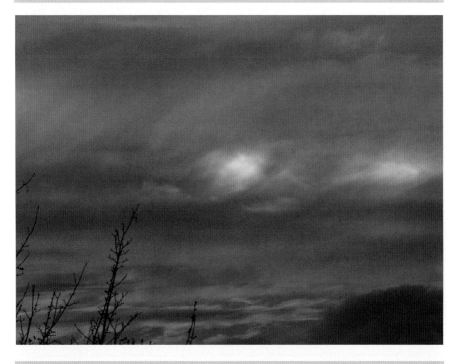

Fig. 8.15. An almost infinite variety of Sun dogs can be observed and photographed.

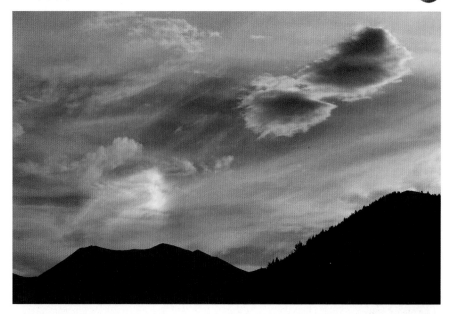

Fig. 8.16. A Sun dog over hills in New Zealand with a little brightness added (credit: Sondela, http://www.sondelart.com).

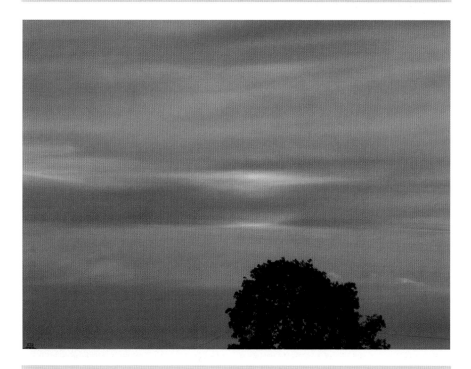

Fig. 8.17. A cloud formation revealing a double Sun dog.

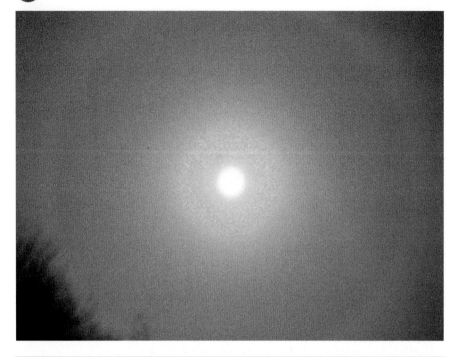

Fig. 8.18. Moon halo too large for the camera field of view.

In contrast to many phenomena that are gone in a flash longer events can result in memorable records – if you are persistent and patient. One August morning, over a period of four hours, the bright Sun and wispy summer clouds provided a spectacle of ever-changing colors from haloes to parhelia and arcs. A real mixture and a great show are presented in the sequence of selected frames in Fig. 8.20. The first alert was a Sun dog among the cirrus clouds. As the Sun appeared over the trees, segments of the halo began to show in ribbons. Another sequence of Sun dogs appeared until, in full sunlight, the complete circle could be seen, although too large to fit into the camera field of view. For the whole four hours the stripes and circles of color hopped from one feathery blanket of cloud to another and were crisscrossed with the usual ubiquitous airplane trails. The show finally ended with another burst of colored arcs and dogs.

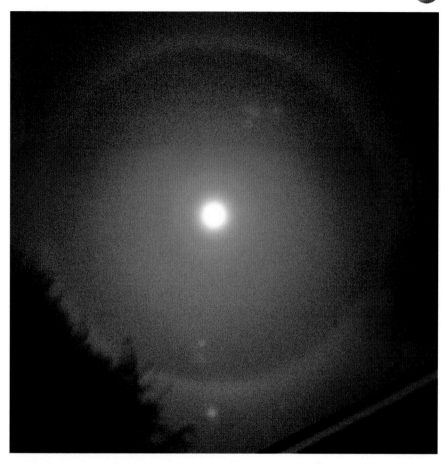

Fig. 8.19. Several frames stitched together to reconstruct the full Moon halo (credit: Greg Smye-Rumsby).

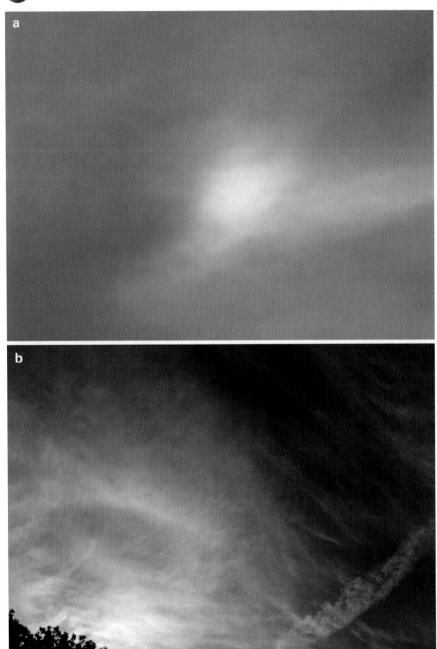

Fig. 8.20. A selection of images from four hours of ever-changing arcs, haloes, and Sun dogs.

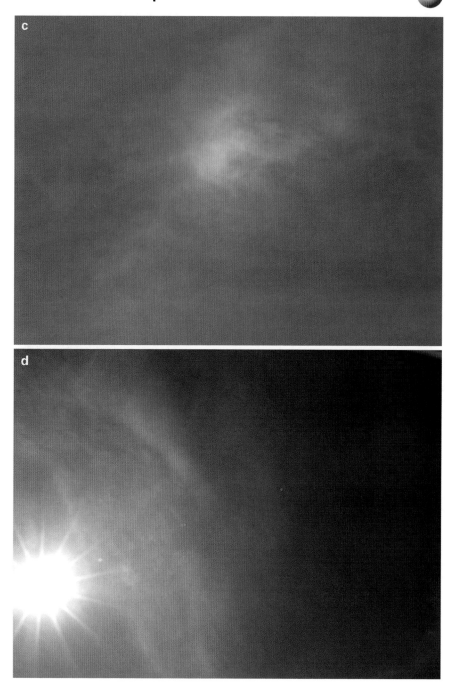

Fig. 8.20. (continued)

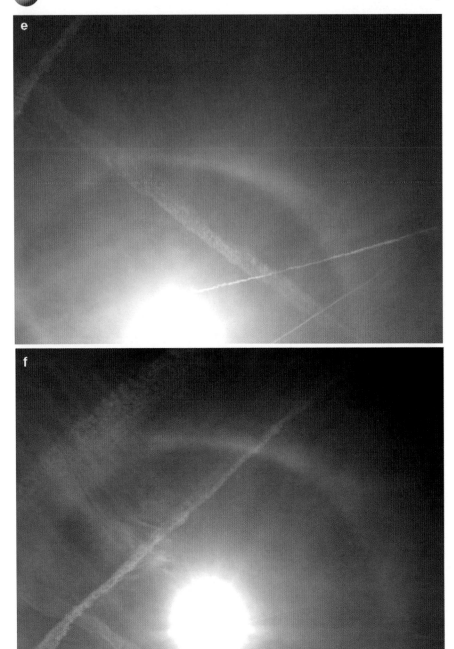

Fig. 8.20. (continued)

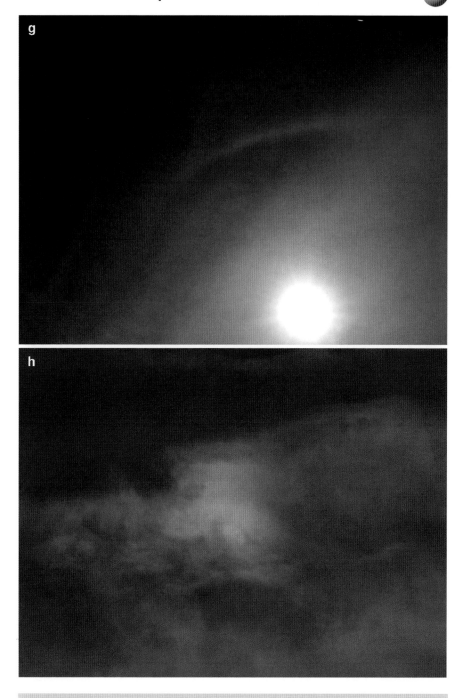

Fig. 8.20. (continued)

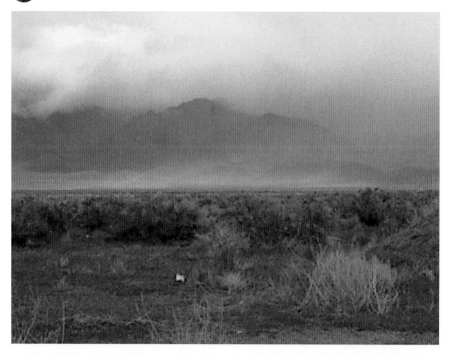

Fig. 8.21. Earthbound rainbow (credit: Judy A. Mosby).

And what about those other bows: starting with a "ground bow" in the mist (see Fig. 8.21)?

Judy Mosby writes, "Two things are odd about this 'rainbow'; it's on the ground, and its colors are weak and widely spaced". It was captured near Lovelock, Nevada, at 4.21 p.m. Pacific Daylight Time on May 7, 2003. At that instant, the Sun was 39° high, and this explains the bow's position. Rainbows are circles of about 42° radius centered on a point directly opposite the Sun. Thus the bow's center was 39° below the horizon, and so the top stood up just 3° above it. The colors? Ordinary rainbows are produced by refraction and reflection of light through large raindrops. Much smaller mist droplets at the base of the mountains formed this bow. Light interference effects in the small droplets broaden the bow and dilute the colors to produce a cloud bow, or when the droplets are very small, an almost white fog bow."

And talking of fogbows, see Fig. 8.22.

Fig. 8.22. Fog bow image captured during a trip to Grand Manan Island, New Brunswick, Canada (credit: Johanne McCullough, http://www.flickr.com/photos/sudsmuffincanada).

Pillars and Rays

The Sun plays great games with the sky and clouds, especially near the horizon. Just as the Sun sinks "below ground" a narrow vertical beam of light reflecting from ice crystals can often be seen, although you have to be quick to see it, as it often lasts for just a few seconds. This is termed a Sun pillar. Once again spot metering is useful here (see Figs. 8.23, 8.24, 8.25). A dramatic Sun pillar is shown in Fig. 8.26.

It is possible to capture a combination of features in one event (see Fig. 8.27).

Dark instead of bright lines above the Sun (see Fig. 8.28) could be termed antipillar (probably just a single crepuscular ray phenomenon – see later). The sky often creates a tapestry out of an event. The fan-like antipillar (see Fig. 8.29) appears to emanate from a volcano drawn in the sky by a pointed cloud.

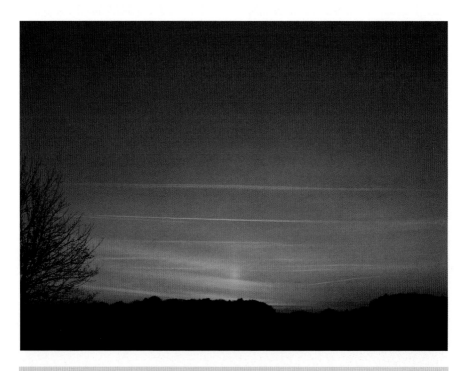

Fig. 8.23. Sun pillars fade very quickly.

Fig. 8.24. Sun pillars fade very quickly.

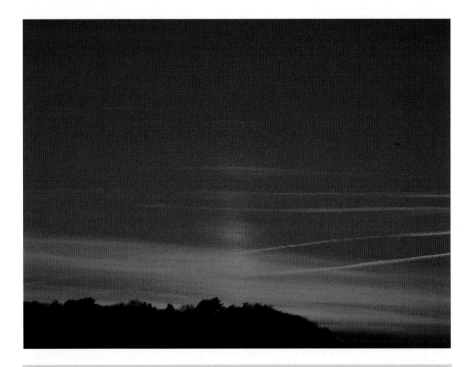

Fig. 8.25. Sun pillars fade very quickly.

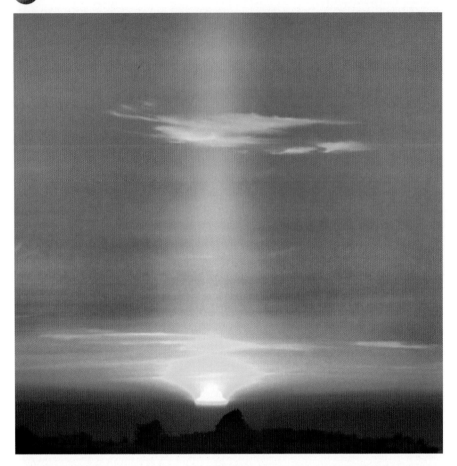

Fig. 8.26. Sun pillar in San Francisco (credit: Mila Zinkova, http://home.comcast.net/~milazinkova/Fogshadow.html).

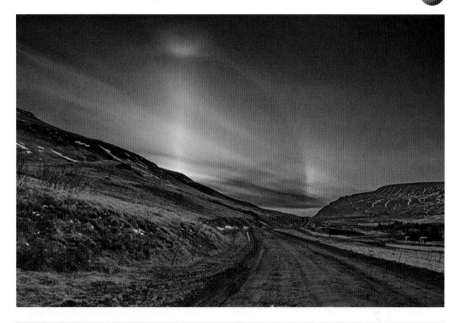

Fig. 8.27. Sun dogs and a Sun pillar illuminate a solar halo at sunset over Iceland (credit: Omar Runolfsson).

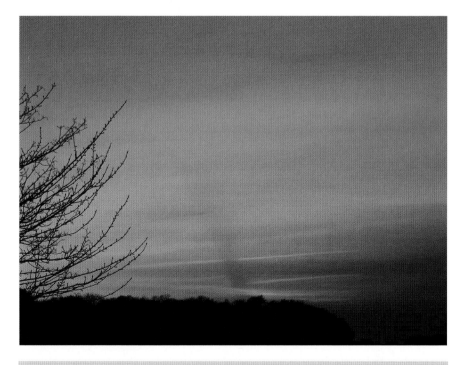

Fig. 8.28. A dark, antipillar ribbon streams up from the horizon.

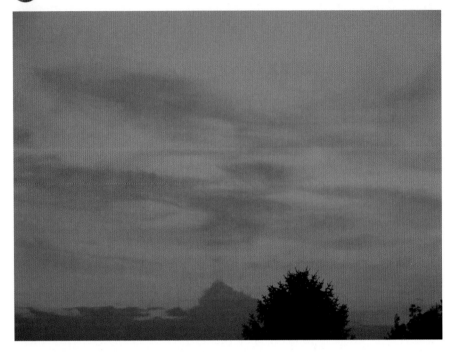

Fig. 8.29. An antipillar volcano drawn in the sky.

What we used to call evening sunbeams many decades ago are more technically referred to as crepuscular rays. They are parallel beams of sunlight made visible due to reflection from particles in the troposphere, the lowest level of Earth's atmosphere. They don't look parallel, though, because of the effect of perspective (see Fig. 8.30).

A little more unusual is the spectacle of rays streaming from the Sun and continuing broadly overhead, possibly to be observed at the opposite horizon as anticrepuscular rays (Fig. 8.31). Anticrepuscular rays can be seen on the horizon with your back to the Sun; the sparkling beams created by the clouds appear to be directed fan-like towards a distant point on the ground (see Fig. 8.32).

Crepusculum is the half-light period between night and day, commonly called "twilight." Crepuscular rays can therefore appear both in the morning and the evening.

Crepuscular rays can also occur underwater, especially in the Arctic, as light shines through cracks in the ice.

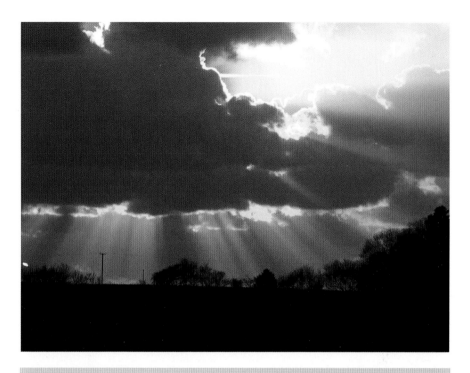

Fig. 8.30. Crepuscular rays streaming onto the distant tree-lined landscape.

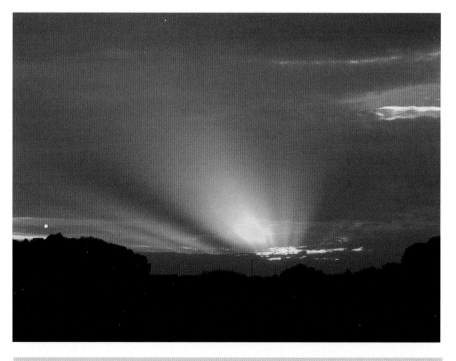

Fig. 8.31. Crepuscular rays streaming out above the clouds.

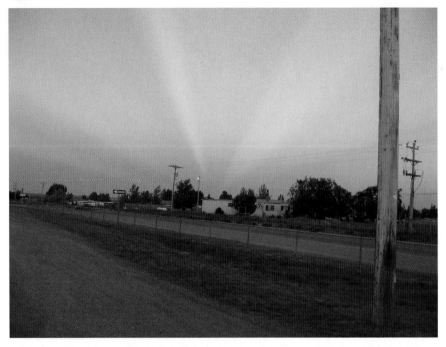

Fig. 8.32. Morning anticrepuscular rays (credit: David and Patsy Kessler).

Because of the difference in the color sensitivities between eyes and camera the Sun's rays can appear even better on a photograph than was actually seen. Crepuscular rays are usually yellow or red, owing to the low angle of the sunlight at these times passing through, and being refracted by, the atmosphere. Many cameras are more sensitive to red and infrared than the human eye. A bright Moon can also provide crepuscular rays or moonbeams, as in Fig. 8.33.

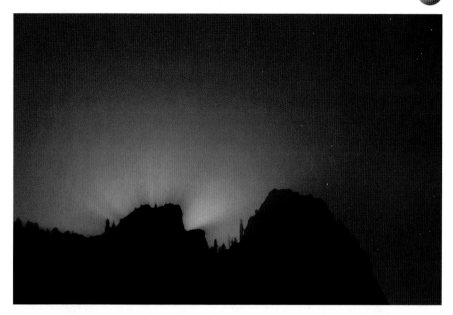

Fig. 8.33. Moonbeams over the hills set a distinctive mood (credit: Mila Zinkova, http://home.comcast.net/~milazinkova/Fogshadow.html).

Clouds

While being alert to the appearance of sky colors, shaped or meaningful clouds that "happen" to pass by also never cease to amaze. What atmospheric parameters led to the circles shown in Figs. 8.34 and 8.35? Just missed the Sun dog that came and went in the center of the first picture!

The Internet makes it possible to get in touch with experts around the globe, and most of them are very kind and cooperative. Members of the UK Meteorological Office kindly provided the following explanation for the "cloud holes": "Precipitation falling from above the cloud where the hole is. The trailing ice streaks (or virga) is the upper cloud version of the rain we get nearer the surface. So when the ice crystals fall into a layer of Altocumulus (for example) below, the moisture in the Ac links up with the falling ice crystals (either bonding if they are both in ice form, or the vapor flowing round the crystal if it is liquid), then continuing towards the surface. Thus the moisture at that spot has been removed, but the surrounding cloud has not been touched, so is left with a hole."

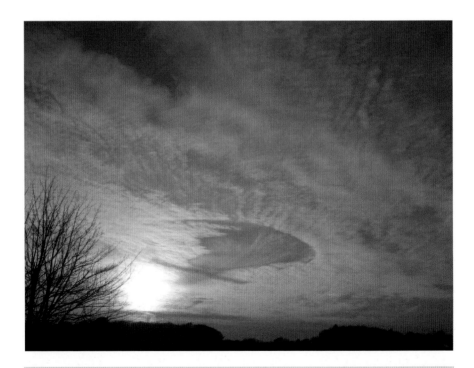

Fig. 8.34. A golden Sun illuminates a circular cloud formation.

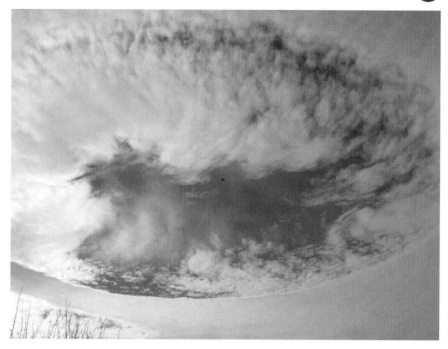

Fig. 8.35. Note the huge blue sky hole in the cloud blanket.

An even more dramatic example of this effect in beautiful blue and white is the image shown in Fig. 8.36, aptly described by the photographers as a "Feather Fall."

The Lowell's write "The photo looks more like a feather than a cloud. It's actually an impressive fall-streak cloud. It was taken March 1, 2008, about noon, above Snowflake, Arizona. While standing in my back yard, our neighbor's son, who was helping us paint, just looked up in the sky and said "Look, there's a feather." I ran in the house, grabbed my camera, and took this photo. Fall-streaks result when the cloud droplets, ice crystals, or super cooled water droplets composing a cloud are agitated in some way and suddenly fall out. Usually, they form when ice crystals from a higher cloud level fall through a lower, thinner cloud layer, as seems to be the case here. Note the clear area of blue sky surrounding the "feather." Super cooled water droplets in this lower layer freeze as the crystals pass through, and this freezing action acts to release heat (latent heat of fusion – not to be confused with thermonuclear fusion), which warms the air and causes the surrounding cloud droplets to evaporate. Thus, this precipitation rarely reaches the surface."

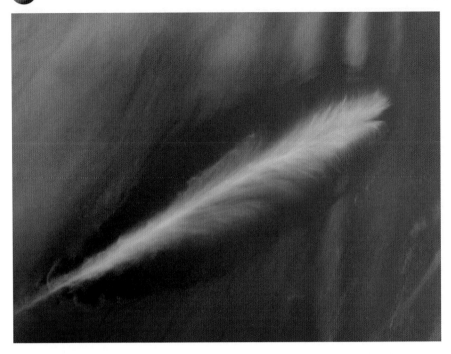

Fig. 8.36. Feather Fall Streak over Snowflake, Arizona (credit: Jerry and Diana Lowell of Snowflake-Taylor, AZ).

An interesting website for an alternative explanation is that of the Cloud Appreciation Society (see the Appendix of this book).

It is worthwhile indulging in one more somber-colored cloud spectacle (see Fig. 8.37), explained by the Meteorological Office as "It could be the top of an old Cumulonimbus (Cb) cloud. I do not know the temperatures involved, but snow appears to be falling from the cloud to the right of the "anvil." So the "anvil" itself could be more precipitation falling from the top of the cloud. I suspect this is the remains of the Cb, the remainder of the cloud having dissipated, leaving the top part of the anvil, which is decaying itself, decaying as its remaining ice crystals fall to Earth."

> When, in the atmosphere, equilibrium develops between the evaporation rate and condensation rate of water, saturation or dew point is reached so the air can hold no more water. If the air temperature is then reduced, water droplets – clouds, will form.

Lenticular clouds can present beautiful and colorful pictures and hover above isolated hills and mountains that are "out on their own." The hills

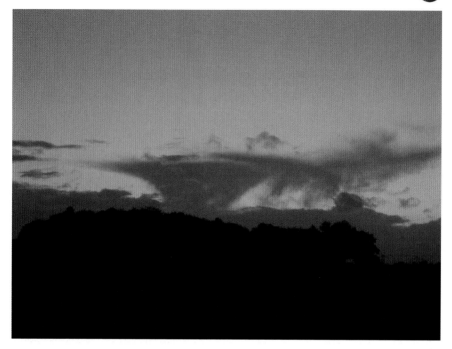

Fig. 8.37. An anvil-shaped cloud.

and mountains force the air that rises over them to cool below the dew point and form clouds. These clouds can remain stationary for long periods of time. Such clouds are frequently lens-shaped, hence lenticular, and develop at right angles to the direction of the wind. Some mountains are almost permanently covered by these clouds, such as those in the Himalayas or covering Table Mountain near Cape Town, South Africa, where they are known locally as the Tablecloth.

> Because of their common disk shape with a bulbous top, lenticular clouds are sometimes reported as alien flying saucers.

Lenticular clouds can be very colorful, with smooth symmetrical shapes. A typical example is shown in Fig. 8.38 over Mount Shasta, California. Mount Shasta rises 3,000 m above a great span of level ground.

Paul Viel writes, "This particular cloud is a layered lenticular cloud. Lenticular clouds, technically known as altocumulus standing lenticularis, are stationary lens-shaped clouds that form at high altitudes, normally aligned at right-angles to the wind direction. Where stable moist air flows over a mountain or a range of mountains, a series of large-scale standing waves

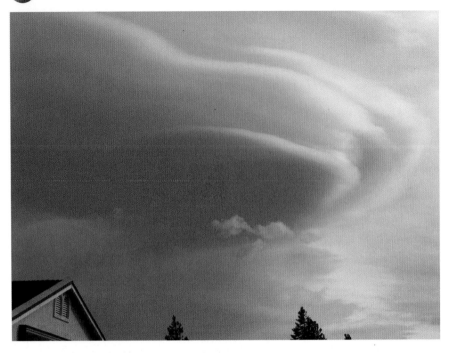

Fig. 8.38. Lenticular cloud near Mt. Shasta, Weed, California (credit: Paul Viel).

may form on the downwind side. Lenticular clouds sometimes form at the crests of these waves. Under certain conditions, long strings of lenticular clouds can form, creating a formation known as a wave cloud." Another prime example is shown in Fig. 8.39. Because of their common disk shape with a bulbous top, lenticular clouds are sometimes reported as alien flying saucers. Typical of the environment of a lenticular, Chateau de Queribus was built on top of the highest peak for miles around.

Polar stratospheric clouds (PSC) occur at high altitudes, up to about 25,000 m, mostly at polar latitudes. Alternatively known as nacreous clouds, they can be quite colorful and of several types, depending on the chemical composition, particularly if they contain nitric or sulfuric acids or water only (see Fig. 8.40). Figure 8.41 shows the eeriness of a Type 2 water PSC.

This stunning photograph (Fig. 8.42) of nacreous clouds was taken on January 25, 2007, near Karuna, Sweden.

A modern, man-made cause of unusual sky colors is the effect of sunlight on frozen water in rocket exhaust. Brilliant iridescence is created by the evenly sized particles mimicking, to some degree, natural nacreous clouds occurring at altitudes of around 20 km.

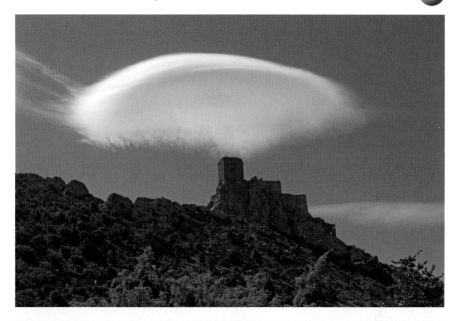

Fig. 8.39. Lenticular cloud over Chateau Queribus, in the Pyrenees, France (credit: Adrian Porter, http://www.Ymzala.net).

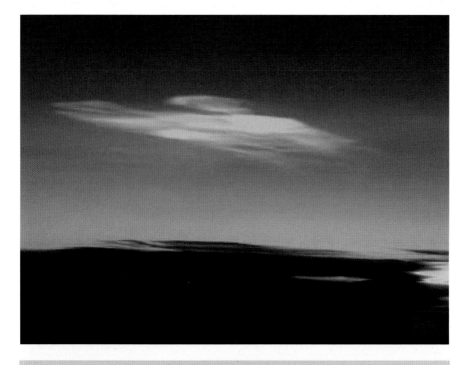

Fig. 8.40. Arctic stratospheric clouds.

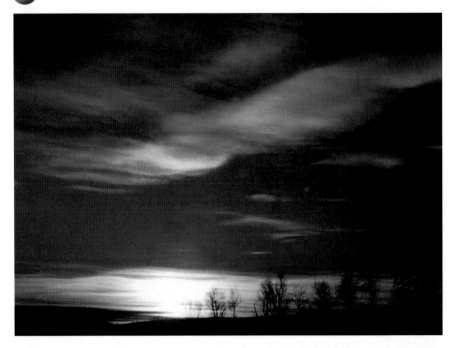

Fig. 8.41. A Type 2 water nacreous cloud (credit: NASA/JPL).

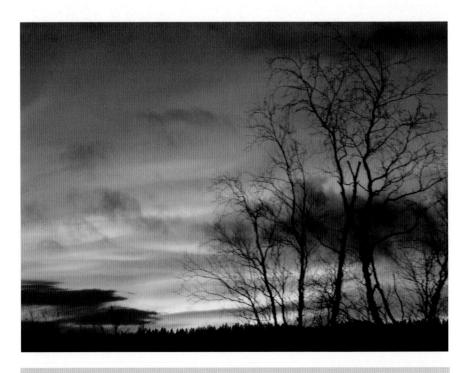

Fig. 8.42. Nacreous clouds over Sweden (credit: Toby Rutland).

Related to nacreous clouds are noctilucent clouds, or night-shining clouds. These are very high, in fact, the highest in the Earth's mesosphere, at around 80 km, where the air is extremely thin. They occur just after sunset in the summer months, when the Sun illuminates the high water–ice (see Fig. 8.43). Noctilucent clouds can be even more stunning when seen over a particularly photogenic landscape (see Fig. 8.44).

> The word "iridescent" derives from Iris, the Greek winged-messenger goddess of the rainbow who flitted between the realms of the gods and the mortal world.

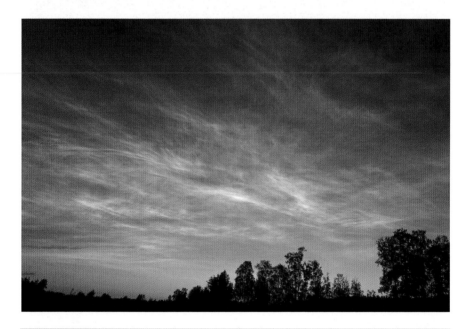

Fig. 8.43. Noctilucent clouds over Saimaa, Finland (credit: Mika Yrjola).

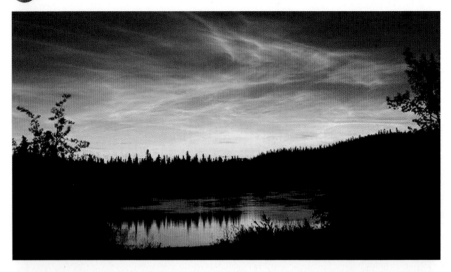

Fig. 8.44. Noctilucent clouds – a sky tapestry (credit: David Cartier).

David Cartier writes, "Noctilucent clouds are a rare and beautiful phenomenon seen only at high latitudes during the autumn, occurring far higher than any other clouds, at altitudes of up to 80 km, where they provide an eerie spectacle by catching the Sun's rays after darkness has cloaked the world below. Never seen during daylight hours, even in perfectly clear skies, they materialize almost magically after darkness falls. Still poorly understood, and until recently thought to be somehow related to concentrations of meteor dust in the upper atmosphere, they continue to be the subject of much scientific research. Aside from all that, they are just an incredible, beautiful experience! I photographed these above the Yukon River near Whitehorse, Yukon Territory, around 2 a.m. on the morning of July 24."

Anyone collecting letters of the alphabet in the sky (and there are such people)? How about a pinkish letter "A" against a pale blue background for a start (see Fig. 8.45)?

Sunsets and Skyglows

Within the ability of anyone with a camera is the capture of stunning sunsets (see Fig. 8.46). They are common and, at their best, are awesome.

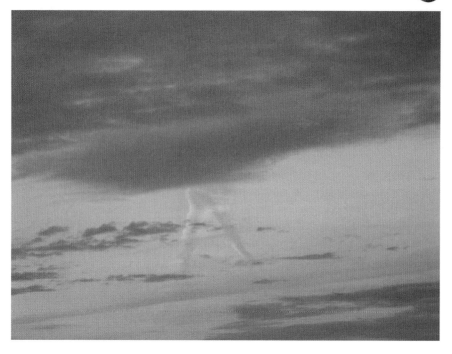

Fig. 8.45. The first of a collection of letters of the alphabet!

The late afternoon blues and purples gradually make way for yellows, golds, even greens, and, finally, stunning reds of all varieties, with their brightness generated by the refraction and scattering properties, Rayleigh Scattering, of the atmosphere and clouds (see Figs. 8.47– 8.52).

A rare event is when there seems to be no cloud covering at all, just an overpowering blanket of deep red, late sunlight reflected from an even spread of particles in the empty sky (see Figs. 8.53 and 8.54).

The UK Meteorological Office provided a likely explanation for this phenomenon:

"Around February 18, 2008, some spectacular sunsets and sunrises were observed across much of the United Kingdom and continental Europe. The sunset colors lasted for up to 60 min after sunset. The general consensus is that they were probably caused by thin PSCs. These form at temperatures below $-78°C$ in the stratosphere, at an altitude of 20–25 km above the surface; on these days temperatures at this altitude were around -87 to $-90°C$." Refer to previous pages for more on PSCs.

While on the subject of effects at sunrise and sunset, there is a phenomenon known as the Green Flash. For just a second or two or even sometimes a fraction of a second a tiny lens-shaped disk appears above the Sun as it disappears below the horizon (see Fig. 8.55).

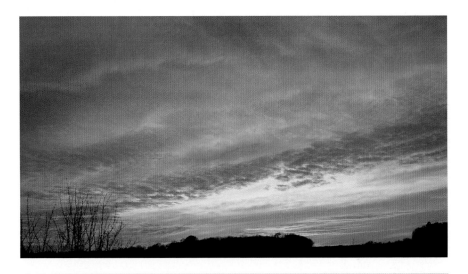

Fig. 8.46. A multicolored sunset.

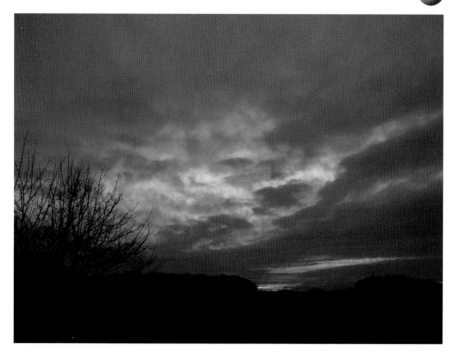

Fig. 8.47. Intense sunset colors.

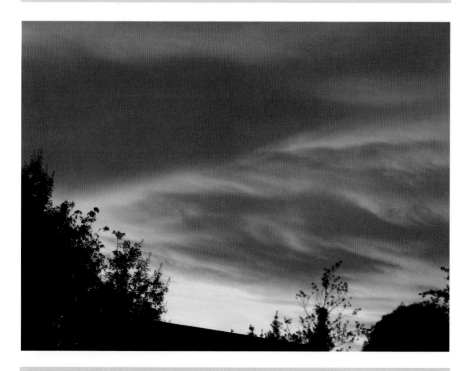

Fig. 8.48. Intense sunset colors.

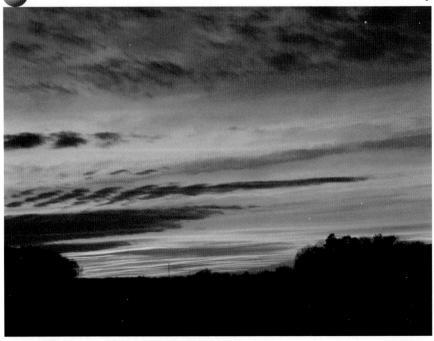

Fig. 8.49. Intense sunset colors.

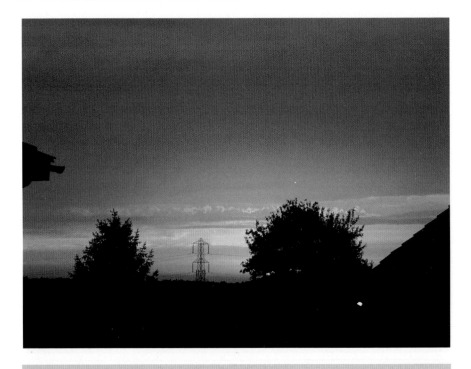

Fig. 8.50. Intense sunset colors.

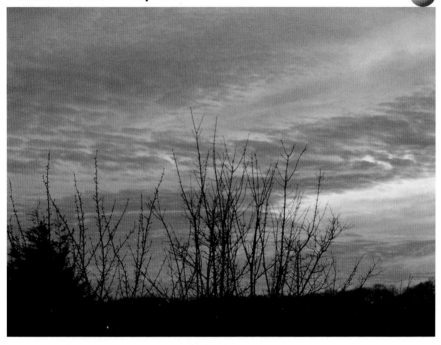

Fig. 8.51. Intense sunset colors.

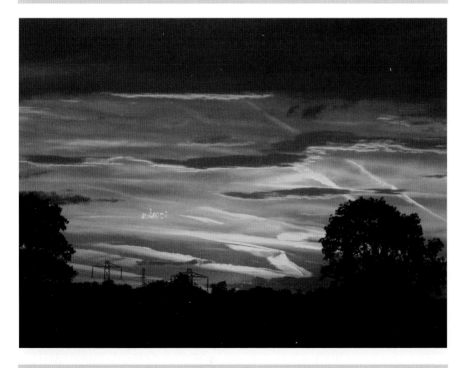

Fig. 8.52. Intense sunset colors.

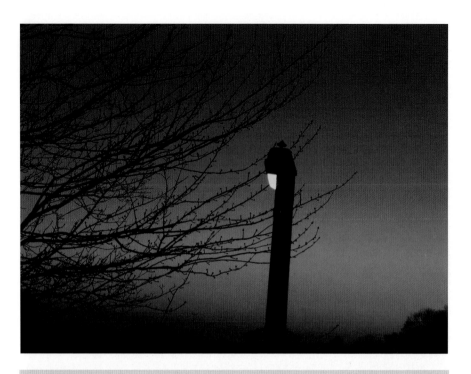

Fig. 8.53. A blanket of deep red from stratospheric clouds covers the evening sky.

Fig. 8.54. A blanket of deep red from stratospheric clouds covers the evening sky.

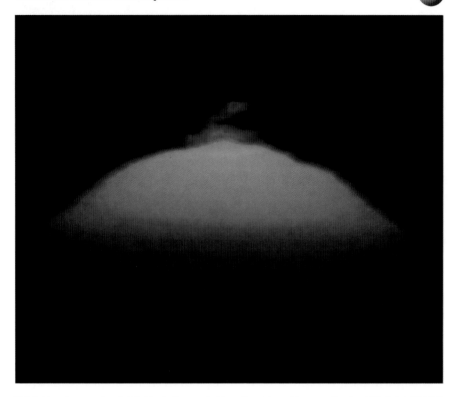

Fig. 8.55. A Green Flash captured as the Sun dipped below the horizon (credit: Mila Zincova, http://home.comcast. net/~milazinkova/Fogshadow.html).

There are several types of this phenomenon and several explanations given in a multitude of references, some of which are incorrect. A reference is given in the Appendix in this book as the most authoritative source, but, essentially, refraction by the atmosphere is, once again, the culprit. Factors include the degree of refraction and the stability and density of layers of air similar in many ways to a mirage seen on Earth.

Just one more sunset from a more exotic place (see Fig. 8.56). Patterson writes,

"The Aguarico River, Sucumbíos, Ecuador, taken from the Cofan settlement at Zabalo. We had spent the day traveling up the blackwater Zabalo River by eco-canoe and had been unusually lucky with the weather. As we returned to the village, the clouds started to roll back in, just as the Sun was starting to hit the tops of the trees. The canoe was a little too shaky for good photography, although it would have made a great vantage-point.

Instead, I waited until we got back and headed for a small clearing on the bank at the most downriver of the village's houses. A little patience paid off with a number of nice shots before fear of tripping over something unpleasant in the dark sent me back.

"The river is a dull muddy brown in daylight, so the orange glow from the sunset definitely adds something. I like the way the Sun lights the edges of the clouds on the right, hinting that it's still daylight in the canopy, even though the darkness under the trees says 'bedtime.'"

Scattering of light (Rayleigh Scattering) is not just an essential ingredient for colorful sunsets. Without it and particles to do the scattering, the sky would be black, just as seen from the Moon. Well, not quite black. Even if we had no atmosphere there are the particles of interplanetary dust that are the source of a permanent but very faint skyglow. This phenomenon, known as the zodiacal light, can be seen in temperate latitudes as a cone of light just after sunset or before sunrise, which is a time of optimum angle with the ecliptic where the densest region of dust resides (see Fig. 8.57). Joe Cali writes, "Camping at Rainbow Valley near Alice Springs. The central Milky Way near Sagittarius and Scorpius hangs over the brilliant zodiacal light. The Zodiacal light is light scattered from interplanetary dust that lies in the plane of the Solar System. For a few hours after the end of twilight in the evening or before the beginning of twilight in the morning, sunlight is forward scattered so that it is easily seen from Earth. To the eye it looks like the light of a small city in the distance and therefore it is best observed from a very dark observing site, where you can be sure no towns are causing the light." At tropical latitudes, in the absence of the Moon and lights, the glow can be seen permanently at night.

John William Strutt, the third Baron Rayleigh (1842–1919), was an English physicist who earned a Nobel Prize for the joint discovery of argon. He was also the first to describe the scattering of light, named after him, and surface waves commonly encountered during earthquakes and other types of seismic activity.

But we *do* have an atmosphere and dust. And particles smaller than the wavelength of light do scatter light. Blue light, having a shorter wavelength, is more effectively diverted than red. As sunlight arrives blue light is scattered and reflected all over the place much more than red, so we see a predominance of blue sky. In the evening red light is more direct, and received, while blue is reflected away; hence, red sunsets and blue daytime sky.

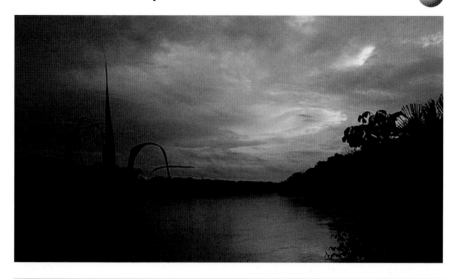

Fig. 8.56. Sunset over the Aguarico River (credit: Anthony Patterson).

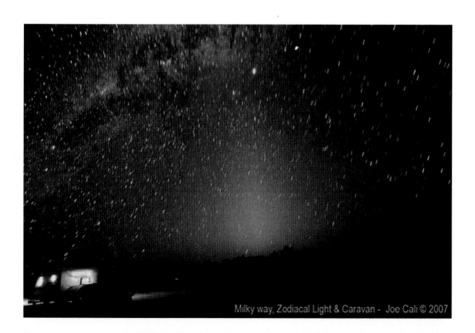

Milky way, Zodiacal Light & Caravan - Joe Cali © 2007

Fig. 8.57. Zodiacal light and the Milky Way. (credit: Joe Cali).

Lightning

At the extremes of the color scale come the white and black. Although professional lightning hunters may spend huge amounts of time, money and effort being at the right time, at the right place, and with the right equipment it is also possible to capture a respectable snap of a streak armed only with a digital compact camera. It is important to know what must be prepared and done when the opportunity of a storm is imminent. Attach the camera to a tripod and be ready to move it to point to any direction that might be the target. Since light pollution, or daylight, could wash out a long time exposure awaiting the vital flash, set to movie mode. As the anticipated time of possibility of lightning arrives, and the area has been determined as extrapolated from constant observation, start the movie and, unless indications are to the contrary, leave the tripod and camera there (Fig. 8.58).

One of the attempts, either this time or another, will lead to capture. Using a movie-editing computer program extract the best frame. As Fig. 8.59, it may not be Earth shattering but very satisfying, and one day... It is worth mentioning the obvious, that you and your camera could get exceedingly wet if the rain actually arrives before the flash opportunity. The example shown was taken while the camera was sited within the kitchen and with the door open! The images presented in Fig. 8.60 are stills taken from over 100 movies obtained during the evening storm of August 6, 2008, over London/Kent, UK.

With the right opportunity stunning photographs of lightning can be captured, especially if other features are involved. Figures 8.61 and 8.62 are two wonderful examples of such images.

On December 11, 2005, near Hemel Hempstead, England, an extremely unusual and huge black layer of clouds drifted by that looked like no other cloud seen before, or likely to originate from a known weather phenomenon (see Fig. 8.63). As news unfolded on that day it became clear that the clouds were the emanations of a disaster. Some 70 or more km away as the crow flies, in Buncefield, a catastrophic explosion of an oil depot occurred, resulting in Europe's biggest peacetime fire. The explosion was reported to have been heard 160 km away and audible in France and the Netherlands. It was measured by the British Geological Survey as a 2.4 on the Richter scale. The dark cloud layers, heavy with the products of combustion, floated by menacingly for the rest of the day, a reminder of the personal tragedies being played out at Buncefield.

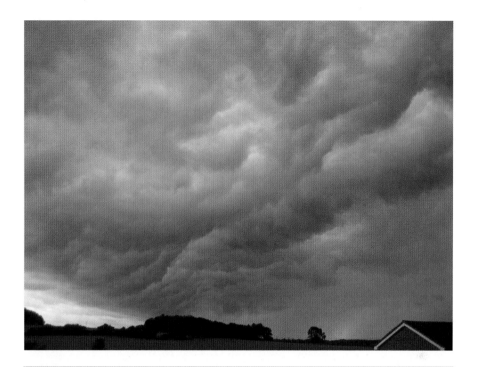

Fig. 8.58. Heavy clouds with following cascades of rain (Mare's Tails) herald a good storm.

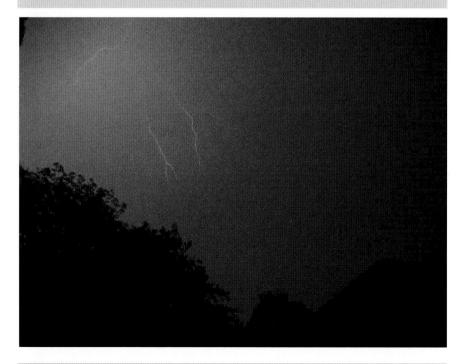

Fig. 8.59. A still photograph belies the power of the atmospheric event.

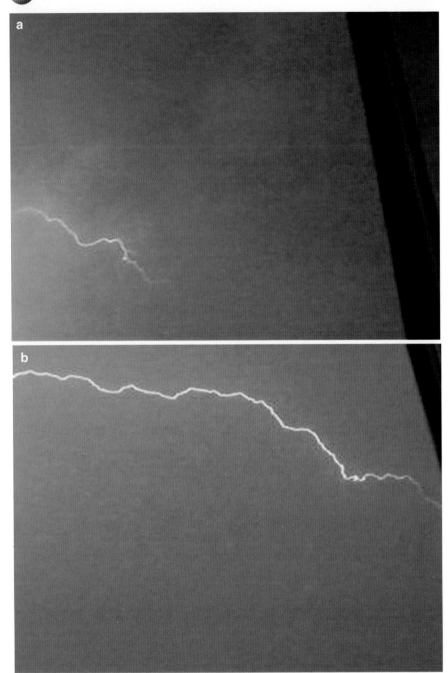

Fig. 8.60. Tortuous trails of lightning illuminate the sky.

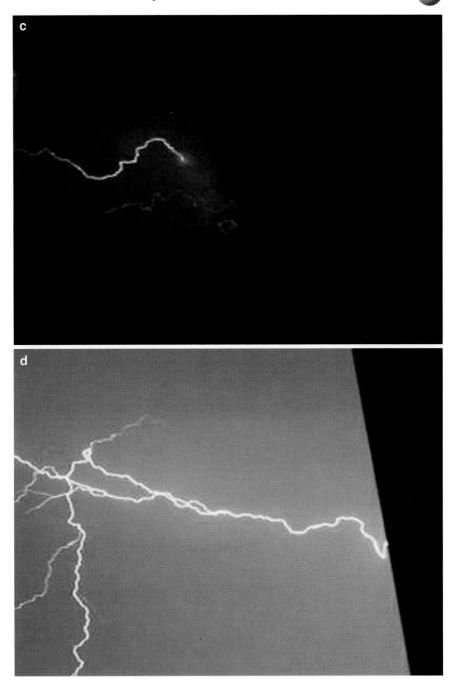

Fig. 8.60. (continued)

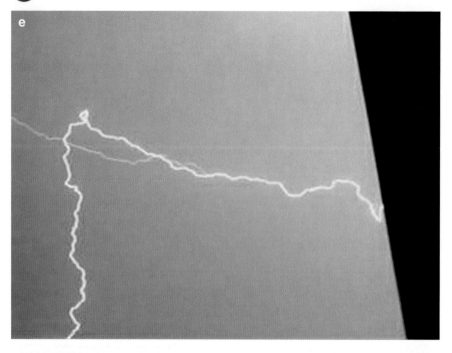

Fig. 8.60. (continued)

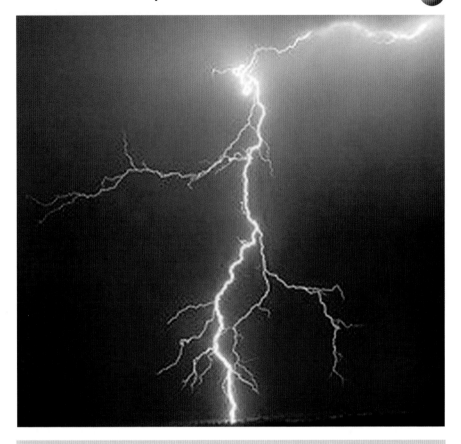

Fig. 8.61. Lightning touchdown in vacant paddocks (I hope) between Meadows and Mount Bold Reservoir, South Australia, December 2006 (credit: Heath Raymond, http://www.redbubble.com/people/Bugalugsrox).

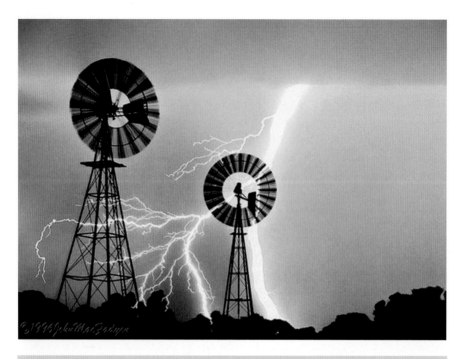

Fig. 8.62. Slide taken during a thunderstorm at Leinster, Western Australia (credit: John MacFadyen, http://www.flickr.com/people/jock49/).

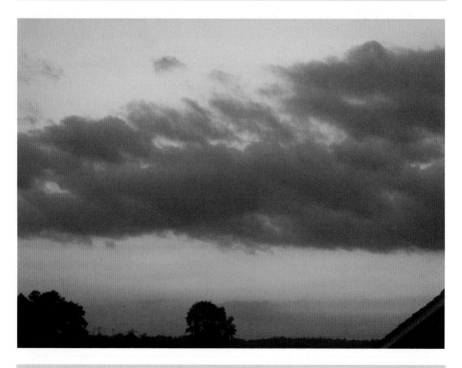

Fig. 8.63. A picture showing the results of an oil depot explosion.

The Moon

A previous chapter introduced the colors of our Moon as seen from space and through large telescopes and a glorious red lunar eclipse using a compact digital camera from the ground. There is more to be seen and collected resulting from atmospheric distortions and interference. As already mentioned a full lunar halo is massive and does not fit within the usual field of view of off-the-shelf cameras, so some frame stitching is needed afterwards on the computer to reconstruct the complete circle. The exact parameters for success at capturing the right brightness and contrast are difficult to predict accurately, so, for such uncertainties, it is best to snap at many settings and dump the failures.

A full or near-full Moon shining through thin lower cloud can create a wonderful display of pastel colors. However, as the hues are not as bright as daytime phenomena the shutter speed must be as long as possible to gather lots of light but as short as possible to avoid unacceptable blur due to the clouds racing across the sky and the glare of the Moon as it skids across a gap in the clouds (see Fig. 8.64). One trick that can often make things easier, or different, is to point the camera *downwards* into a puddle of water (see Fig. 8.65).

The Moon's reflection can appear more colorful on camera than as seen owing to the eye's (monochromatic) rods being more responsible for night vision than the cones (see Fig. 8.66).

Once in a Blue Moon really only refers to a second full Moon within the same calendar month, but other colors can be seen. During a particularly nasty spread of forest fires in the United States the picture (Fig. 8.67) of a smoky Moon was captured by Nikolai Sklobovsky. Below that are some additional images of the Moon and Sun (Figs. 8.68 and 8.69).

Color by accident is commonplace in astronomy. Attempting to capture Jupiter over the River Blackwater in County Cork, Ireland, resulted in wonderful artistry of colorful reflections (Fig. 8.70).

There are moments when the magic and beauty of the Moon is displayed as if it is a personal communication just to you. This reflection across the bay from the Youghall peninsular in Ireland says "Everything will be alright. I'll always be there for you", Fig. 8.71.

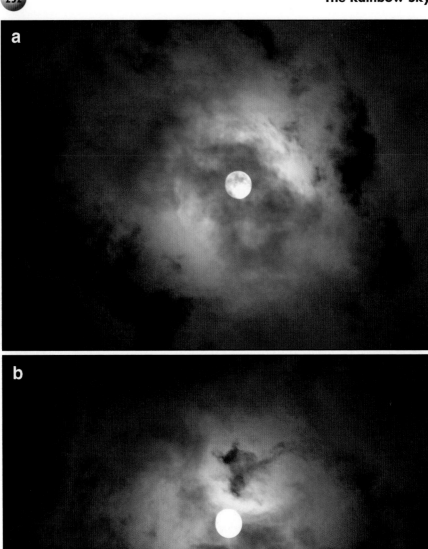

Fig. 8.64. Moon colors caused by the refraction of the Sun's reflected light through moisture in the clouds.

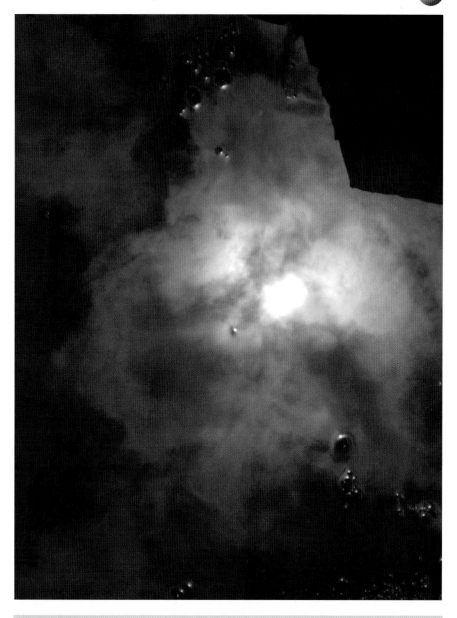

Fig. 8.65. Reflections from a puddle.

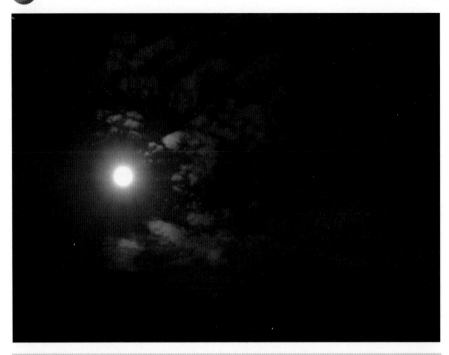

Fig. 8.66. Colors from a full Moon. The slight blurring is due to the necessity for a longer time exposure of 4 s.

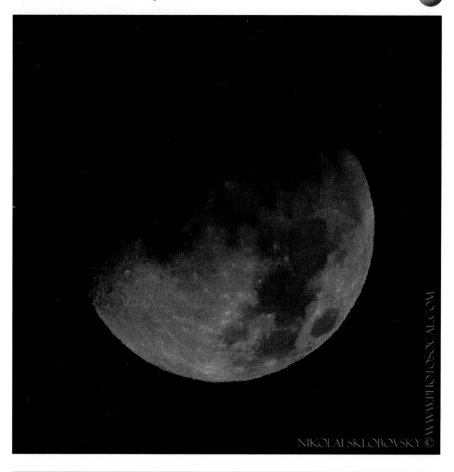

Fig. 8.67. Smoke from a forest fire gives the Moon an unusual color. (credit: Nikolai Sklobovsky).

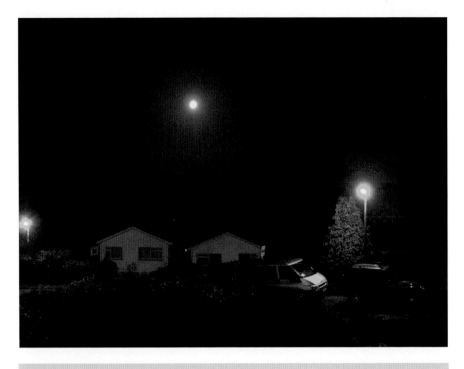

Fig. 8.68. Full Moon over a country village.

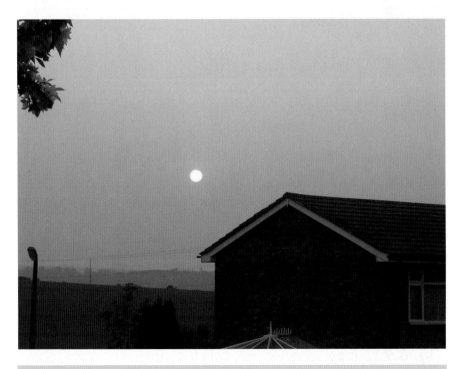

Fig. 8.69. A hazy evening Sun.

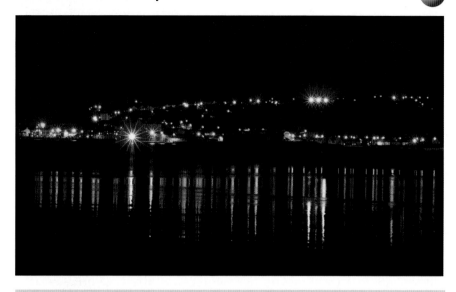

Fig. 8.70. Colorful seafront reflections.

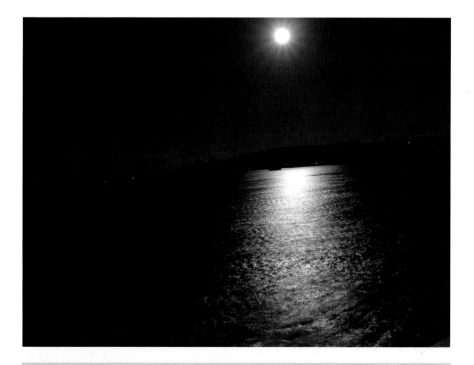

Fig. 8.71. A personal moment with the Moon.

Colors from Above

All these ephemeral and colorful images are presented to us while we are at ground level. But take some time out from reading, dozing, or being bored when taking an airplane trip somewhere. Look out of the window as far below the airplane as possible. A ring of color may be seen when the Sun is shining from directly above and a layer of cloud is below. The phenomenon is called a glory. Although the ring can be as beautiful and colorful as a good rainbow, photographs do not do it justice due to the window material and the angle of viewing through it (see Fig. 8.72). The picture shows the airplane silhouette nicely circled by the usual sunlight spectral colors.

The picture is a little faint, so the use of computer enhancement is helpful to learn more about the finer details. Figure 8.73 is the result of adjustments to a combination of brightness, contrast, gamma, saturation, and hue effects to provide much extra information on the structure of the glory. If the airplane is very high above the cloud the airplane's shadow may be too small to be seen within the glory.

Notice that there is not just the one ring visible from the airplane. There is clearly a repeated pattern of concentric rings of color. Notice other aspects, too, such as the order of the colors and the position of the plane. Because the glory is due to an alignment of the plane, the Sun, and a portion of cloud – a straight line through them – the camera (you) is precisely in the center of the circle. So it is apparent exactly where you were sitting! That is probably not of great consequence (unless you should not have been sitting there!) but interesting and a contributing fact when considering the origin of glories. As the airplane reduces altitude during preparation for landing and gets closer to the cloud, its shadow gets bigger and bigger until it swamps the whole bottom half of the glory. Figure 8.74 shows a computer-enhanced image of one stage of the reduction of distance between plane and cloud. This alternative combination of computer enhancements also brings out the feature's fine structure.

Because glories are opposite the Sun, at the antisolar point, they are always below the horizon and cannot be seen except partially at sunrise or sunset – or from a high place such as a mountain or from a boat above sea mist.

If a low Sun shines from behind a person to produce a glory the distorted image can appear very ghostly and is merely the shadow of the person instead of that of the airplane already described. Such an apparition is called a Brockenspecter and, as for anticrepuscular rays, perspective may cause the image to appear pointed or triangular, making it even more scary (see Fig. 8.75).

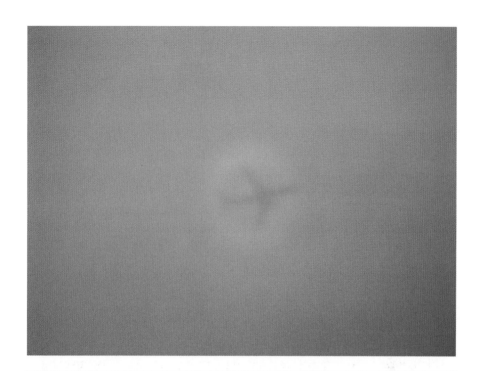

Fig. 8.72. A glory image captured through the window of a Boeing 737.

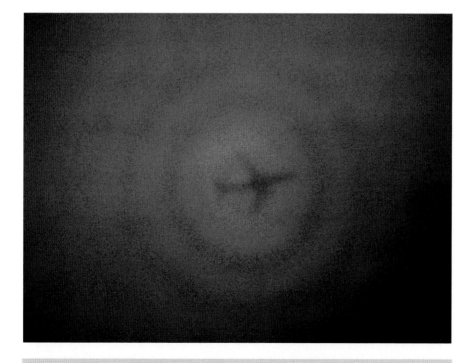

Fig. 8.73. A computer edited image of Fig. 8.72.

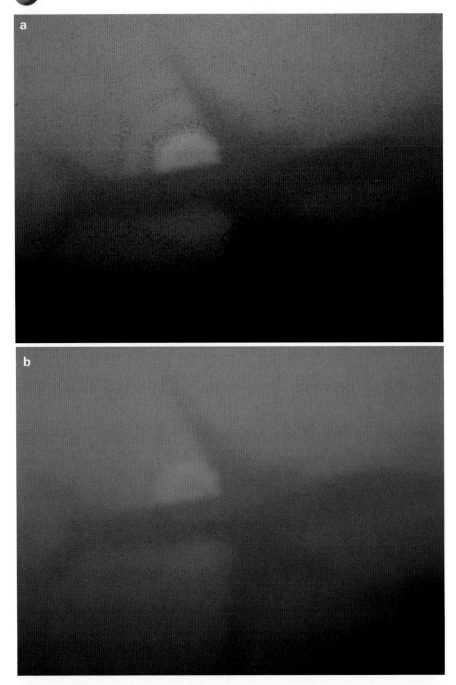

Fig. 8.74. Computer-enhanced image of an airplane on glory as the plane comes in to land.

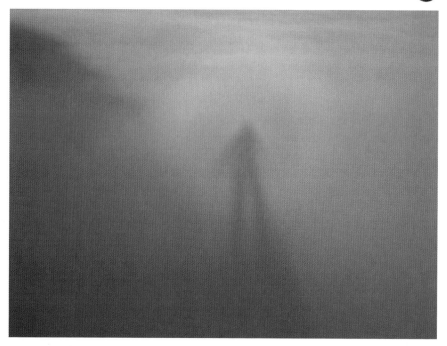

Fig. 8.75. A Brockenspecter on High Crag (credit: David Fitzgerald)

David Fitzgerald writes, "I took the photograph in January 2006 while walking on the High Stile range in the Lake District (UK) during the best and most extensive cloud inversion I have ever experienced. On my descent I realized that the chances of a Brockenspecter were quite high, so I waited, just above cloud level, for about an hour while the Sun lowered in the sky and was rewarded with this super photo."

Are the colors of glories produced in the same way as rainbows or Sun dogs? No! Although it is generally acknowledged that the precise explanation for the glory is unknown, there are in-depth descriptions of successful simulations, for example, the Mie theory and the Debye series reformulation of the Mie series. Reference to this is given in the Appendix of this book, but in summary, the colored rings of the glory are caused by two-ray interference between "short" and "long" path surface waves, which are generated by rays entering the droplets at diametrically opposite points.

Ghostly shadows, called Heiligenschein, can also be created by light reflected straight back from water droplets hanging on leaves or leaf hairs.

This is another antisolar phenomenon, so the shadow of the photographer will be hauntingly projected onto the backdrop.

Yet another antisolar phenomenon is the opposition effect. Imagine an uneven surface such as a layer of sand, gravel, plant material, etc. with the Sun directly above. All particles will have a shadow, and observing from the side there will be some shadows and some reflections. For views from directly above, the amount of shadow seen will be minimal, as the shadows are now below the particle, thus giving an overall greater average brightness than views from the side. This can manifest itself as a bright halo around the projected shadow of the observer.

The opposition effect is often cited as the explanation as to why the full Moon is so much brighter (relatively speaking) than the Moon at other phases, although there are many factors that contribute to the glow, such as retroreflection by crystalline minerals and a phenomenon known as coherent backscattering. The label "opposition effect" is derived from the fact that it was first discussed as being mainly responsible for the brightness of the Moon and Mars at opposition. Figures 8.76 and 8.77 show

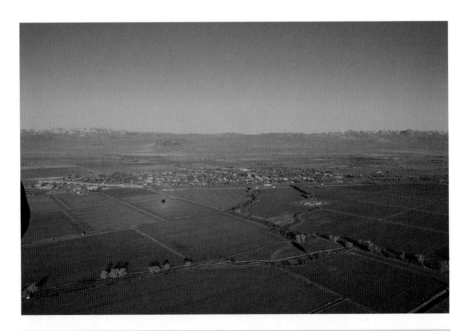

Fig. 8.76. Opposition effect as seen from a balloon ride (credit: Judy A Mosby).

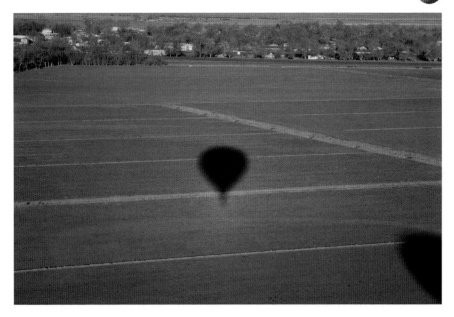

Fig. 8.77. Opposition effect as seen from a balloon ride (credit: Judy A Mosby).

the effect seen from a hot air balloon. A bright spot traveling along the ground below an aircraft is another example of this effect. The parachute of the Mars Exploration Rover Opportunity cast a shadow as it floated down to the surface of Meridiani Planum, and surrounding the shadow a bright halo was spotted, again, because of the opposition effect.

Aurora

No collection of atmospheric colors would be complete without the eerie silent phenomenon of an aurora (see Figs. 8.78 and 8.79).

Aurora Borealis, the Northern Lights, and Aurora Australis, the Southern Lights, are created when highly energetic particles from the Sun cascade into the upper atmosphere, causing the atoms and molecules to become excited and emit light as described in a previous chapter. The particles are accelerated along the magnetic field lines of the Earth's magnetosphere toward the poles.

EISCAT (European Incoherent Scatter Scientific Association) is a radar observatory situated in the cold arctic regions of Sweden, Norway, and Finland that is dedicated to studying the Earth's upper atmosphere, the home of the Northern Lights. It has discovered that the most ionized part of the ionosphere that stretches from around 100 to 1,000 km is about 240 km high, 60 km less than previously thought. The sensitivity of the instrument is staggering, being able to detect a cloud of electrons the size of a pinhead at 400 km and space debris of a couple of cm at 1,200 km. By combining EISCAT's data with that from satellites and ground-based telescopes much more will soon be discovered and understood about these awesome spectacles.

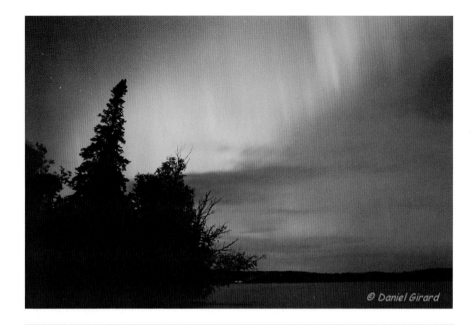

Fig. 8.78. Stunning aurora images captured over the Laurentides wildlife Reserve (Lac des Ilets), Québec, Canada (Credit: Daniel Girard, http://www.spacew.com/gallery/image002833-2.html).

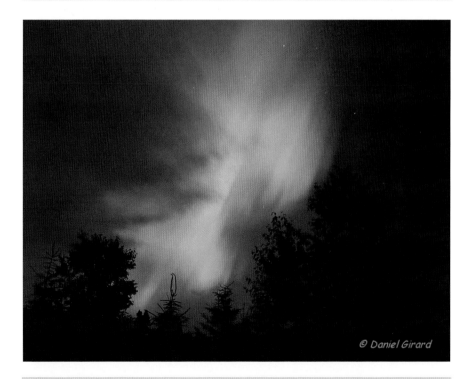

Fig. 8.79. Stunning aurora images captured over the Laurentides wildlife Reserve (Lac des Ilets), Québec, Canada (Credit: Daniel Girard, http://www.spacew.com/gallery/image002833-2.html).

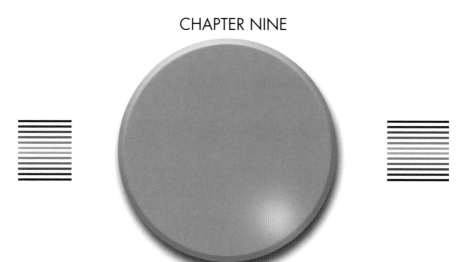

CHAPTER NINE

Man-made Colors in the Sky

For a considerable time complex life forms have been able to produce light. Recent discoveries have revealed the bioluminescence emitted from the awesome creatures deep within the oceans. But it is not necessary to delve into the abyss to witness such phenomena. Phosphorescent plankton illuminate the surface of the sea at night, and glow worms and fireflies flit about making their modest contribution (*see* Fig. 9.1).

Not until modern humans came onto the scene did serious and constant brightening of the landscape arrive; early human's cooking fires hardly count. Much of humankind's contribution to colors in the lower troposphere comes under the umbrella of light pollution, often referred to as skyglow – not to be confused with the natural phenomenon of zodiacal light (*see* Fig. 9.2). Many astronomy societies and organizations are spearheading awareness of the problems that light pollution causes. An excellent example is shown in Fig. 9.3.

Some darkness must be preserved in order for our children and future generations to be able to stare in wonder at the sparkling gems in the night sky. A great way to assess the visibility is to select a patch of sky and count the number of stars that can be seen within. Repeating the exercise at various locations gives a good idea as to the geographical extent of pollution, although seeing conditions must be taken into account to ensure comparing like with like. Joining in such national projects can make a

T. Buick, *The Rainbow Sky*, Patrick Moore's Practical Astronomy Series,
DOI 10.1007/978-1-4419-1053-0_9, © Springer Science + Business Media, LLC 2010

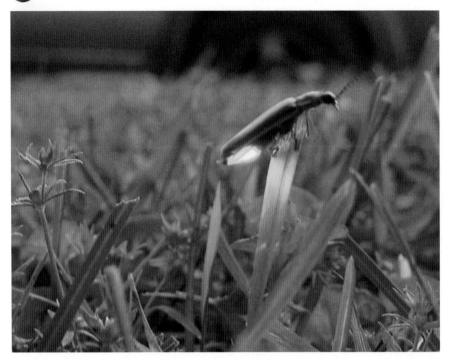

Fig. 9.1. Firefly glowing (Credit: Scott Craig).

significant contribution to the collection of data by those who can make a difference. However, the colors are there, and not all of them are harmful or permanent. Even a simple view of a seafront dwelling captured with a compact camera (when actually looking to capture Jupiter) can look beautiful; for example (*see* Fig. 8.70). Colored lasers enhance the spectacle for visitors to the Niagara Falls (*see* Fig. 9.4).

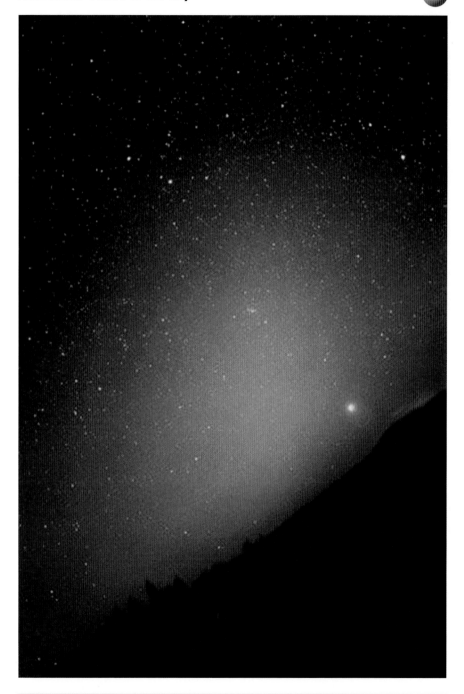

Fig. 9.2. The zodiacal light passing through the constellation of Cancer (Credit: Dominic Cantin).

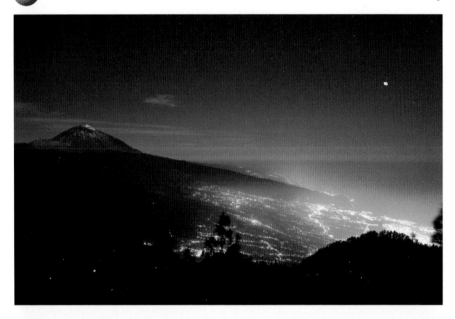

Fig. 9.3. Light pollution in north of Tenerife (Credit: José Àngel Estévez, http://www.cestomano.com).

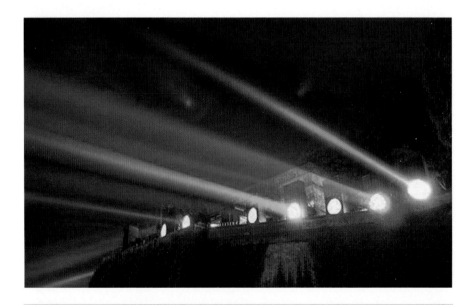

Fig. 9.4. Lighting up Niagara Falls, Ontario, Canada (Credit: Mark Visosky, flickr.com/photos/mvisosky/2773084852).

Satellites

One of the most amazing achievements of humankind is the creation and launch into space of the International Space Station. From the sophisticated technology and the bravery and skill of those prepared to risk the potential dangers many lessons will be learned, essential to prepare for the next human steps into the harsh environment of interplanetary space. The huge solar cell panels reflect sunlight, and the bright white spot can be seen, in a clear sky, speeding its way across your sky in a minute or two and circling the world in around 95 min per orbit.

Very acceptable pictures with evening hues can be formed by including some trees or other background during capture. The time of appearance and trajectory across the sky can be forecast by visiting an appropriate website (*see* the Appendix of this book) and plugging in coordinates applicable to your location. Note the path as exactly as possible, which is made easier by noting any prominent stars or constellations in the line of flight. On a tripod, point the camera at your chosen target area and set to 30 s exposure. Experiment with different times to find the best for you and the extent of light glare in your area (*see* Fig. 9.5). With practice it is also possible to capture a second interval and a third, etc., by quickly moving and aiming the camera to achieve a record of start, maximum, and finish of the complete pass (*see* Fig. 9.6).

Planning to capture images of the bright ISS spot near objects is rewarding (*see* Fig. 9.7). Also thrilling is to observe the tiny bright spot as it transits the Sun and the Moon, as shown in Fig. 9.8. A really skillful astronomer might swing the telescope across the sky to capture more detail of the ISS structure, as in Fig. 9.9. It can be quite addictive collecting more and more images of trails while remembering the technological achievements and the courage of the astronauts represented by that tiny white spot.

Now and then, something unexpected and dramatic shakes our senses and sets our pulse racing. Under that heading, for most, fall the rings of Saturn when seen for the first time through a good-sized telescope and the plethora of unique features sculpted on the surface of our Moon. A chance look up above to see for the first time a bright spot, brighter than any night sky object other than the Sun or Moon, can be another startling moment. Many are first compelled to assign the phenomenon to a UFO sighting, but, not only is there a rational explanation, its precise arrival can be predicted. It can even be seen flashing overhead in the daytime sky. It is, of course, the reflection of the Sun's light from the solar panels of a satellite, an Iridium satellite that, when overhead, can briefly present

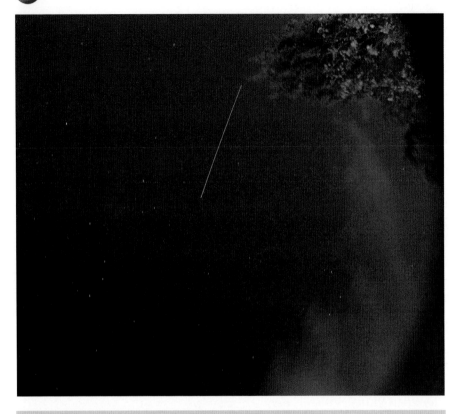

Fig. 9.5. It is possible to photograph the ISS trail even under a light-polluted sky.

a flare of magnitude up to around −8 or −9, over 30 times brighter than Venus at its brightest. The three highly reflective door-sized aluminum panels are covered with silver-coated Teflon for thermal control.

There are occasions when the reflected light from a satellite can be photographed when close to the horizon, at which position noticeable refraction occurs as shown in Fig. 9.10. Much more satisfying is catching a magnitude of −8 almost directly overhead (*see* Fig. 9.11).

Most astonishing of all is the dedication and precision adopted by enthusiasts who chase double Iridium flares. It is challenging enough to capture a single, but photographing a double arrival is great and is even better with a landmark included, as in Fig. 9.12.

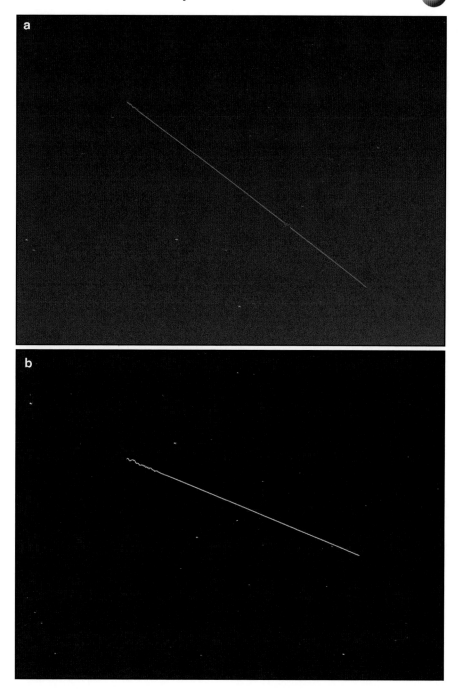

Fig. 9.6. A sequence following the complete track the ISS from west to east.

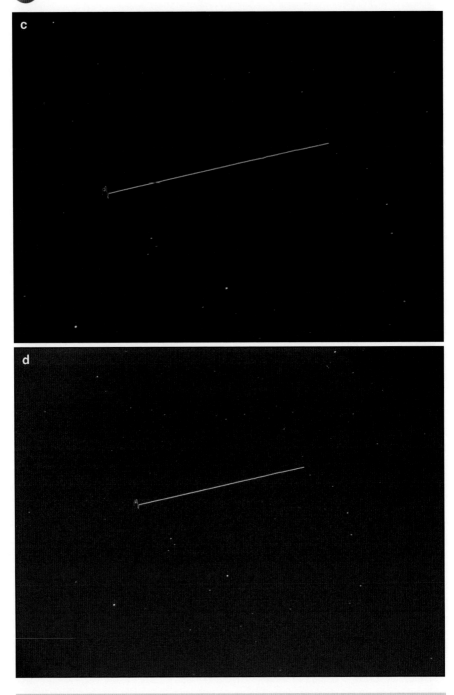

Fig. 9.6. (continued)

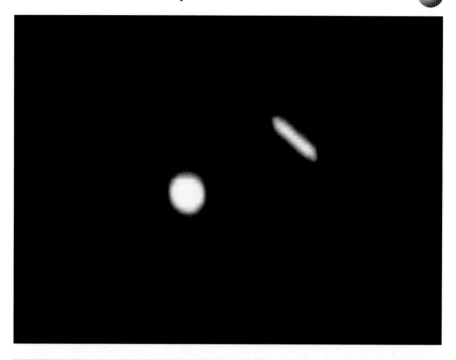

Fig. 9.7. A still from a movie of the ISS speeding past Jupiter.

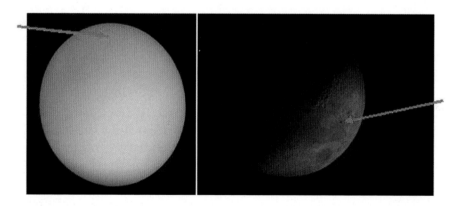

Fig. 9.8. The tiny dark spot can just be seen as it transits the Sun and the Moon.

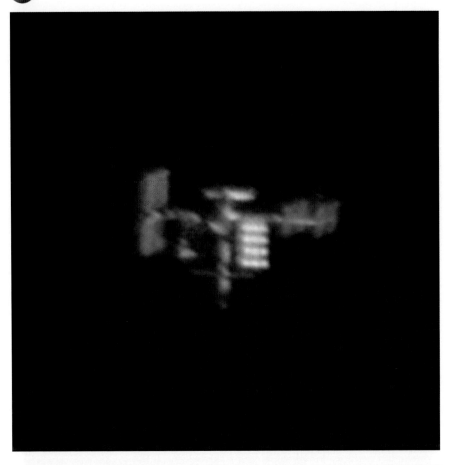

Fig. 9.9. An amazingly detailed image of the ISS (Credit: Mark Shelley, Orpington Astronomical Society).

Many Iridium satellites are constantly being renewed to maintain a system of global communication to serve handheld phones, especially useful for remote locations such as oceans and polar regions. Originally planned were 77, the atomic number of Iridium.

Chris White Writes of His Photograph:

On February 7, 2007, I was made aware of an upcoming dual Iridium flare with almost no delay between the two flares and both center-line's within a half mile of each other just east of my location in the Metro Minneapolis/St. Paul area via the Minnesota Astronomical Society. I had figured that the optimal spot between the center lines

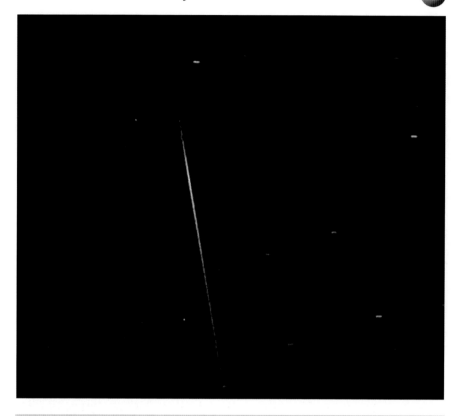

Fig. 9.10. A distant Iridium satellite flare, a flash of light at maximum sunlight reflection.

was along 93.1705 W (Dec Deg NAD27), that the flare would occur at about 7:23:33 P.M., give or take a second every 2 miles north or south using a combination of Orbitron and Heavens Above. I then began going up and down the line on a map looking for suitable foreground objects but was unable to select one. I plotted where the flares would occur in the sky using Cartes Du Ciel by creating a Star Chart with the location of the flare plotted with red finder circles that would enable me to 'frame' the shot.

While you could easily use a compass and degrees above the horizon, using the stars to frame your shot is like using graph paper to compose your shot. In this instance the flare occurred directly on the line of the two bright stars in Ursa Major ('The Big Dipper') that point to Polaris (the 'North Star'). After an hour or two of planning over a couple

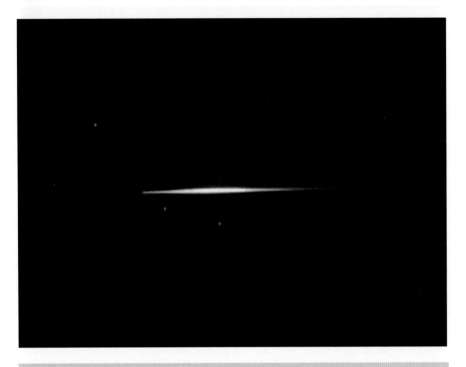

Fig. 9.11. Iridium 70 at its best.

days, I had still not found a suitable foreground object for the dual Iridium flare. The temperature was around −1°F with 5 mph winds that evening, so I did not want to do any hiking. I finally selected my target at about 6 P.M. that evening, the Church of St. Peter in Mendota Heights, MN. I left my house around 6:20 P.M., finding the church at about 7:10 P.M. I took 10 min setting up the camera and captured the image at around 7:23:31 P.M. The flares from Iridium 59 and 96 were expected to be in the −8 range (−7.9 and −8.4 between centerlines along 93.1705 W), occurring nearly simultaneously, which they did. At 7:22 P.M., I looked up in the sky and started counting to 90. I first saw the flares like dim fireflies and hit the shutter release. The flares grew to incredible intensity then back to fireflies before disappearing under the urban light-polluted skies 14 s later.

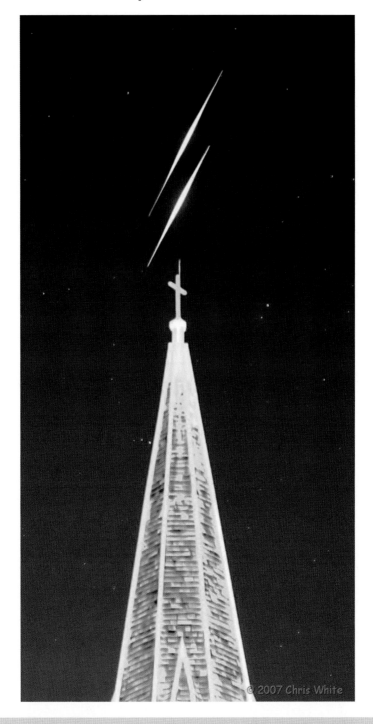

Fig. 9.12. Double Iridium flare over the Church of St Peter (Credit: Chris White, http://www.thesuckzone.com).

Fireworks

Calling the events of the early universe "fireworks" would be the grossest understatement ever, but that is just how many refer to recent discoveries by NASA's Spitzer Space Telescope. It is strongly suggested that infrared light detected in a prior study originated from clumps of the very first objects of the universe more than 13 billion light years away. Massive stars, possibly 1,000 times larger than our Sun, were exploding and spitting out huge amounts of energy. The information was gleaned by masking out stars, galaxies, and other objects to produce the "firework image" below (*see* Fig. 9.13).

It is a good thing that nature's current grandest explosions in space are so far away even though they would be an astounding spectacle. The gamma ray burst, GRB, from a massive star at the end of its life would annihilate anything nearby. Theia, a Mars-sized object, crashing into early Earth to create the Moon would also have been fun to watch – from a safe distance. Fireworks are the best we can really do in our back garden or at a public display (*see* Fig. 9.14). The Catherine Wheels are like galaxies and rockets are like meteors. The colors are wonderful but gone in a flash.

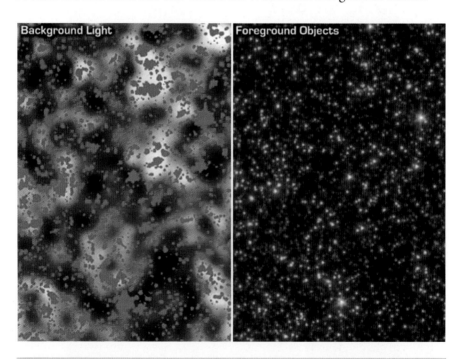

Fig. 9.13. First Light after the Universe's dark ages (Credit: NASA/JPL-Caltech/A Kashlinsky (GSFC)).

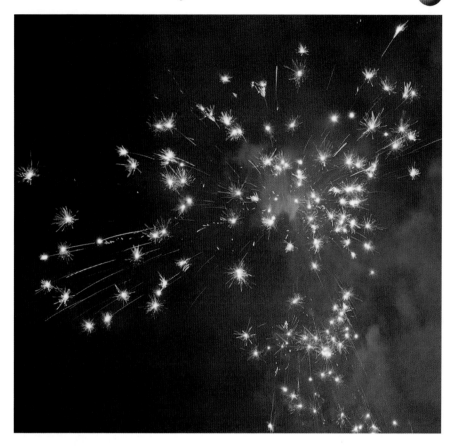

Fig. 9.14. An explosion of stars.

There are many different theories on the origin of fireworks, but most agree that gunpowder was probably first discovered 2,000 years ago by the Chinese when a cook dropped saltpeter (potassium nitrate), charcoal, and sulfur on a fire by mistake. Varying the proportions of the three chemicals produced different shades of colors and different levels of brightness when compressed into bamboo tubes and exploded. Firecrackers, supposedly invented by a Chinese monk named Li Tian about a 1,000 years ago, became popular because they were thought to drive evil spirits away with the lights and noise generated. Sparks were a good sign and the smoke was a healthy atmosphere.

Marco Polo or the Crusaders may have brought back the gunpowder to the West, where it was soon used for military purposes. Queen Elizabeth the

First of England so enjoyed the spectacle of fireworks displays (a recorded display in August 1572, for example) that she created the position of "Fire Master of England." Li Tian resided in the Liu Yang region of Hunan Province, where most of the world's fireworks are manufactured today, although they are also produced in Spain, Germany, and South America.

Fireworks are no longer made in England. It is interesting to note that firework displays became very popular towards the latter part of the sixteenth century, but, for a while in the seventeenth century, possession of fireworks was declared to be illegal. This is of no surprise since, in 1605, Guido Fawkes and his accomplices placed about a ton of black powder in the coal cellars under the main parliamentary building on November 5th. King James I was, at that very time, attending the state opening of parliament. The plot would almost certainly have been successful had it not been for one of the accomplices having second thoughts and "grassing" in a letter to some members of parliament.

Some of the well-known fireworks manufacturers have been in business for hundreds of years, for example Brock and Pain, although company takeovers during the twentieth century have seen the rise of new manufacturers. Indeed, John Pain established Pain Fireworks around 1593 near Bow Bridge in East London and is reported to have been the supplier to Fawkes in 1605.

Taking photographs of the speedy, flaming particles can be disappointing, with blurs or under- or over-exposure. However, there is a technique that is almost always successful for single shot capture of fireworks, as opposed to a continuous, low resolution, movie presentation. But first, let's consider the science behind the production of fireworks, which has much relevance to the colors observed in stars.

> The construction of the Royal Observatory at Greenwich was paid for by the sale of surplus stocks of military powder during the reign of King Charles II.

The creation of fireworks is a complex business, especially in terms of making them artistic and safe. The basic ingredients are an oxygen producer – a fuel – something to hold the materials together (a binder), and a color producer.

There are two mechanisms employed for color, incandescence and luminescence. Incandescence is when heat generates the color. The hotter the material the more the visible color develops from red to orange, then yellow and white. Control of the burning dictates the color of the charcoal (for example), and the ignition of metals such as magnesium, aluminum, and titanium can further increase the temperature. To produce an almost infinite range of colors requires energy changes within the atoms of each material leading to luminescence.

The promotion of electrons within an atom to a higher energy state, and their return to the ground or another energy state, leads to the absorption or emission of light. The wavelength of that light (the energy of the photon) determines the color as described in a previous chapter. The mechanisms (laws of physics) are just the same as for stars. Elements have their own characteristic colors when heated or "burned." For instance, sodium (Fig. 9.15)

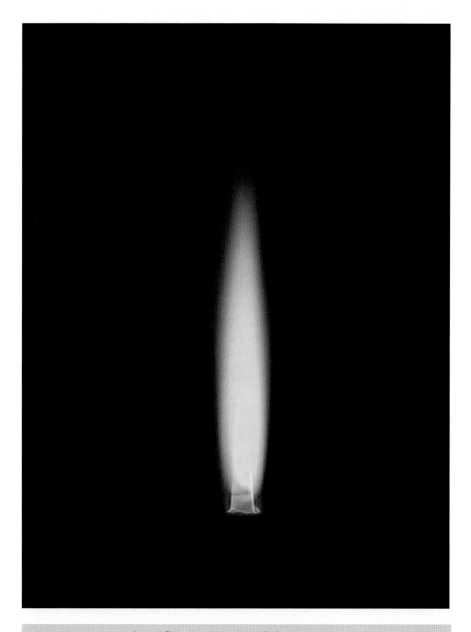

Fig. 9.15. Sodium flame (Credit: Philip Evans).

has a yellow-orange color but is such a strong color that it can mask others, even in tiny quantities. Table 9.1 lists a few characteristic colors of commonly used elements; the table fill colors are only approximate for guidance.

The precise color can depend on the compounds used and other factors. Many will recall identifying in a chemistry lab the chemicals and elements in a simple mixture. Introducing a piece of platinum wire soaked in a solution of the material being tested, then putting it into a Bunsen flame would reveal a color that could be compared with a list, leading to identification. Flame-test colors are given in Fig. 9.16. The element color identifications are: *center,* copper; *top left,* calcium; *top right,* caesium; *lower left,* potassium; *lower right,* strontium.

Perfection of the combustion or explosion process leading to fireworks, warfare munitions, and explosions for peaceful purposes owes much to the development and understanding of modern physical chemistry. In particular, we are talking about the second law of thermodynamics, which addresses the consequences of change in entropy, a measure of disorder, and the laws of heat content (enthalpy) and changes in free energy. More information relevant to the physical chemistry of fireworks is given in a reference in the Appendix section of this book.

So what about those fireworks pictures? The best distance is about 100–200 m, just right for photographing a local display. Set the digital camera on a steady tripod to automatic exposure. Point the camera towards the next expected sky area – not too difficult if watching a small show. Operate the shutter on an auto setting, and "infinity." Under a dark sky it will not close until the brilliance of the fireworks causes an immediate reaction that simulates your catching the event at just the right time. More often than not, at least one out of many pictures taken will be stunning and exhibit the intricacies of the chemical reactions and particle

Table 9.1.

Color	Compound or metal
Red	Strontium and lithium
Orange	Calcium
Yellow	Sodium
Green	Barium and copper
Blue	Copper
Gold	Heating iron and charcoal
Brilliant white	Very hot magnesium or aluminum
Silver	Burning magnesium or aluminum
Purple	Mixture of strontium and copper

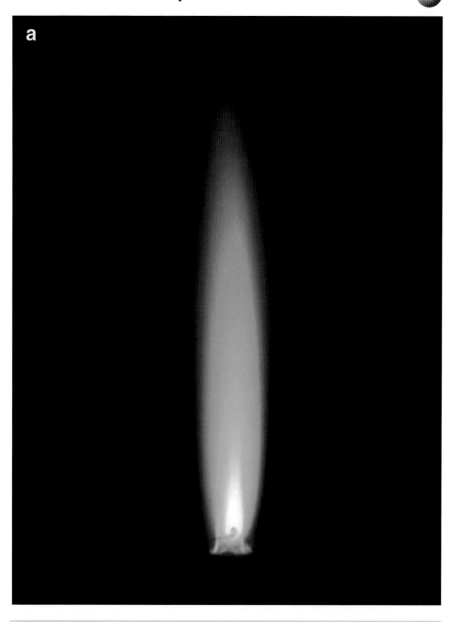

Fig. 9.16. Flame-test colors (Credit: Philip Evans).

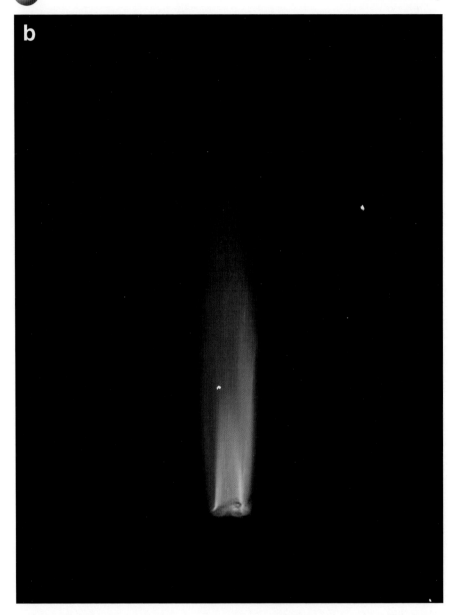

Fig. 9.16. (continued)

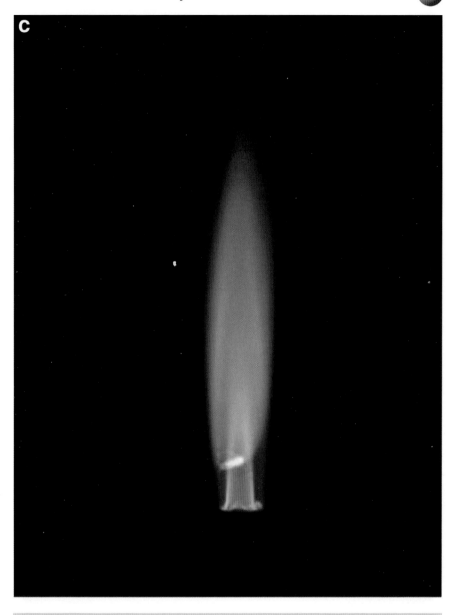

Fig. 9.16. (continued)

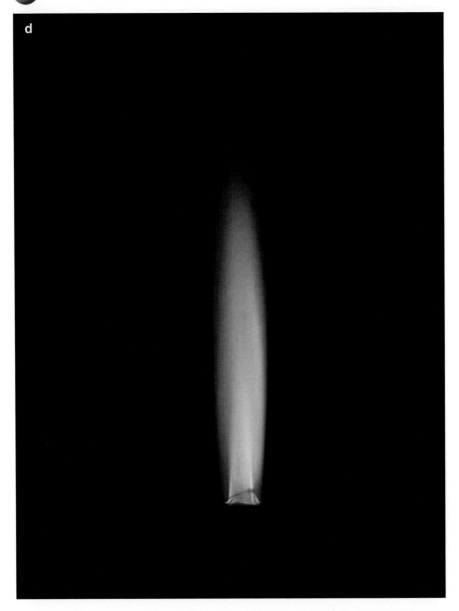

Fig. 9.16. (continued)

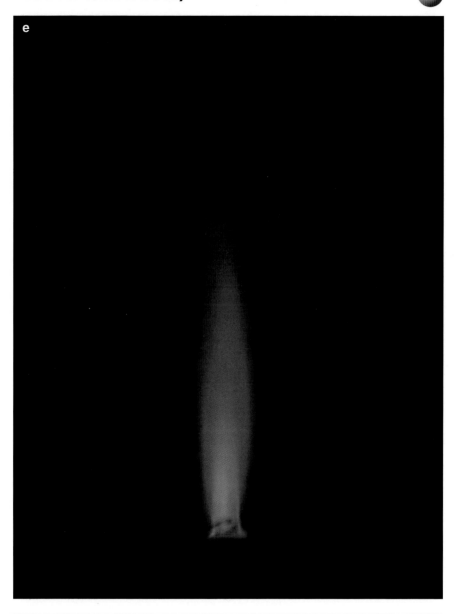

Fig. 9.16. (continued)

flight paths, as in Fig. 9.17. Figure 9.18 shows a colorful picture, but the display is spoiled by the use of cheap fireworks that contain poor ingredients, with the resulting cloud of smoke.

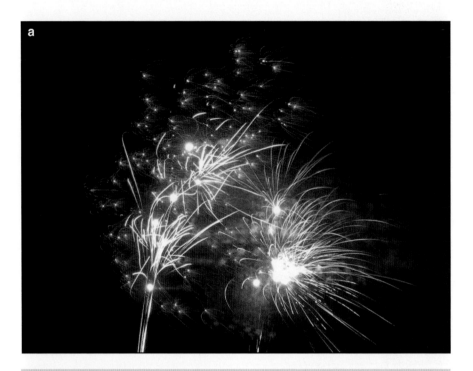

Fig. 9.17. Star-spangled colors and trails at a New Year's celebration.

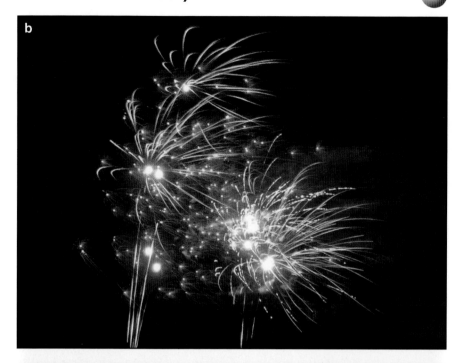

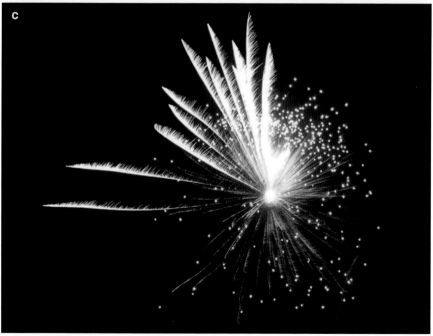

Fig. 9.17. (continued)

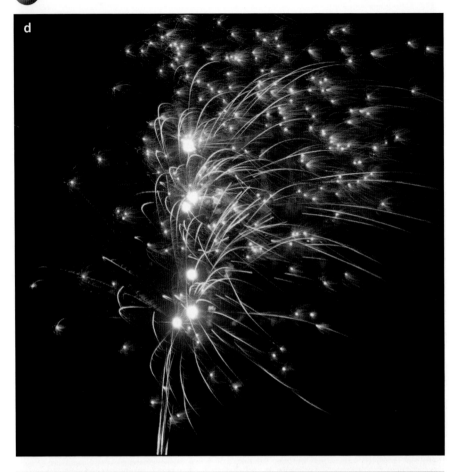

Fig. 9.17. (continued)

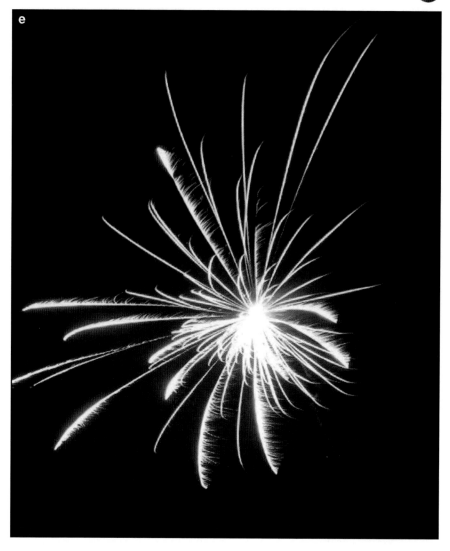

Fig. 9.17. (continued)

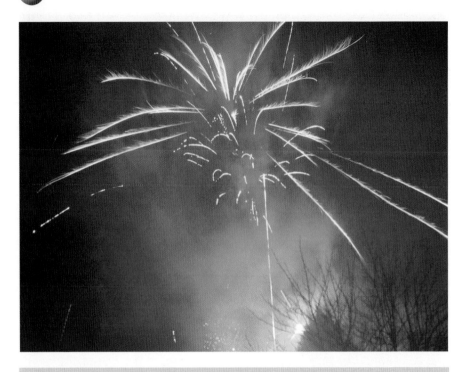

Fig. 9.18. Firework color reflecting from its own cloud of smoke.

Lasers

Laser shows can be wondrous displays of modern technology or infuriating for the light pollution they cause. (LASER is an acronym for Light Amplification by Stimulated Emission of Radiation.) Laser pointers can be helpful and are readily available for use as sky pointers for astronomers and leveling devices for builders as well as for weapons' applications. They are used in medical and dental surgery, CD players, barcode readers, cutting and welding tools, spectrometers, and many other things.

Whereas ordinary light from a torch produces light of many different and random wavelengths (colors) that easily diverge, a laser emits one wavelength. Remember from a previous chapter that to produce light an electron must move from one atomic energy level to another. That happens in all of the stimulated atoms, but not all at the same time. It is not a coordinated emission. Laser light is coherent, which means that the electrons causing the emission of a photon in every atom are stimulated to jump energy levels at the same time. For instance, a flash of light directed into a ruby rod excites the ruby atoms.

Some of the photons produced bounce back and forth between mirrors, exciting more atoms and creating more light of the same wavelength. The monochromatic light can escape through one of the mirrors that is half-silvered thus generating a collimated (traveling parallel to the axis of the rod), single phase, very powerful light.

There are many types of lasers, which can be solid, liquid, or gas and emit different wavelengths. Gas lasers using helium and helium/neon are common and emit visible red light. Dye lasers employ complex organic dyes that can be adjusted to emit light over a wide range of wavelengths. A laser show over the water at Kuala Lumpur in November 2007 presented a super photographic opportunity (*see* Fig. 9.19).

A famous landmark laser is the single, narrow, green light beaming over the Thames from the London Greenwich Observatory on the meridian and is beautiful for what it represents, 0° longitude, about which so much has been written concerning the history of navigation at sea (*see* Fig. 9.20).

The meridian was established by Sir George Airy in 1851 and formally agreed upon as the international Prime Meridian in 1884. Also of note is the fact that a GPS receiver places 0° around 102.5 m to the east of the actual meridian, following an agreement that it should be represented by the mean of star observations (the geodetic meridian) in different countries (*see* Figs. 9.21 and 9.22). Owing to the movement of the Eurasian tectonic plate the astronomical meridian is drifting eastward towards the

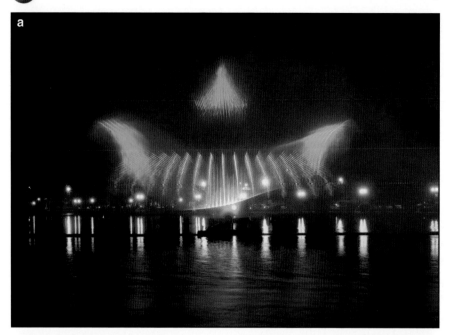

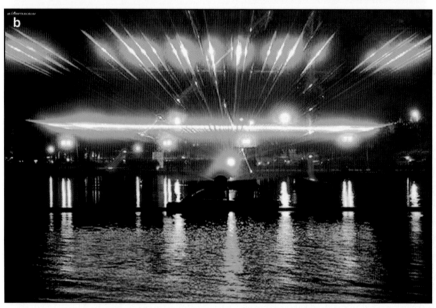

Fig. 9.19. Colorful display of laser lights over water (Credit: Samhaan Hameed).

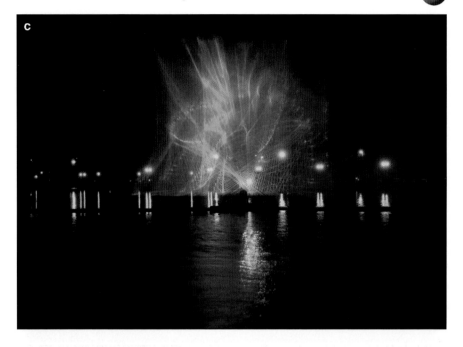

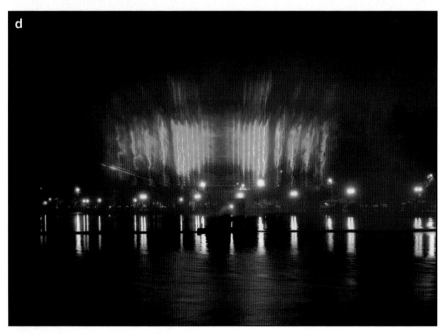

Fig. 9.19. (continued)

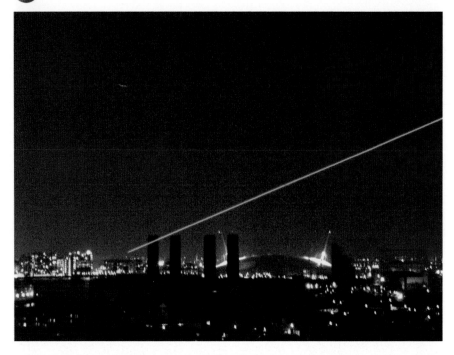

Fig. 9.20. A green laser marking the meridian position.

geodetic meridian at about 1 cm per year. A fitting image to represent the interaction between people and the universe can be beautifully simulated in a planetarium (*see* Fig. 9.23).

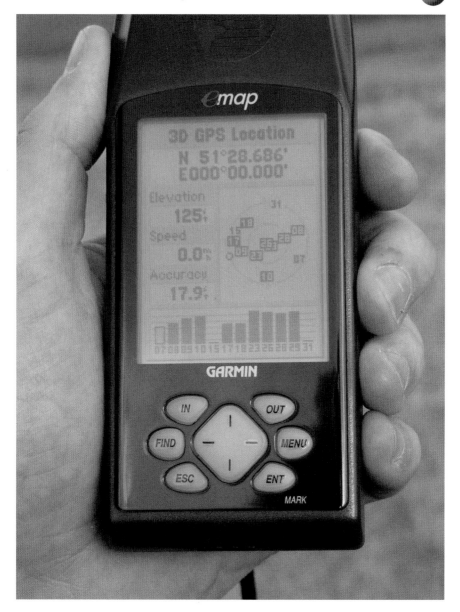

Figs. 9.21. A GPS receiver indicates 0° longitude some way from the Greenwich Observatory, which houses the meridian line (Credit: Chris Buick).

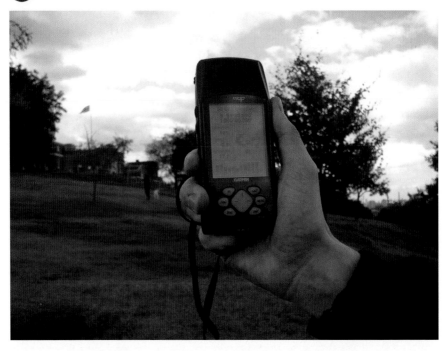

Figs. 9.22. A GPS receiver indicates 0° longitude some way from the Greenwich Observatory, which houses the meridian line (Credit: Chris Buick).

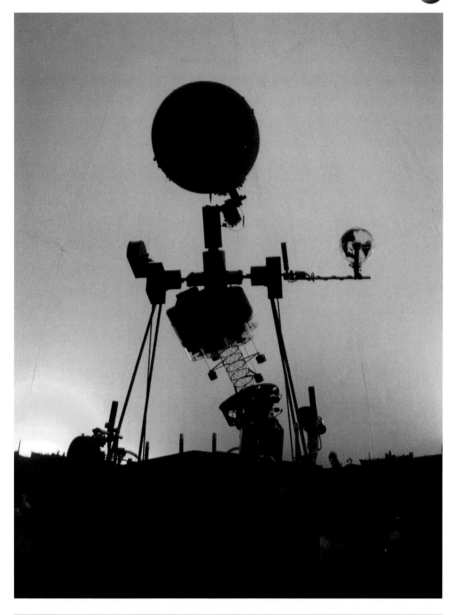

Fig. 9.23. A planetarium projector silhouetted against a sunset backdrop.

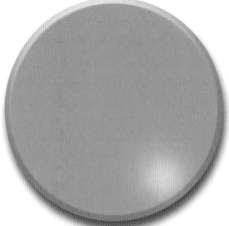

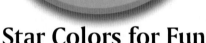

CHAPTER TEN

Star Colors for Fun

Look up at night and, weather permitting, many stars can be seen to fill the vast canopy (*see* Fig.10.1). As we have noted, Hipparchus, around 120 B.C., cataloged 850 stars, rating their brightness on a scale of 1–6, 1 being the brightest. Table 10.1 shows the brightness of some familiar stars.

Most stars appear as bright white spots because the eye is very poor at distinguishing colors in point sources, a condition known as small field tritanopia. But there certainly are some stars that are perceptibly different. Betelgeuse in the constellation of Orion is clearly more orange than white, an indication of its "coolness." At various times of the year other stars are known to present a tinge of orange or red, for example Arcturus (Alpha Boötis) and Aldebaran (Alpha Tauri).

The longer it is possible to stand in the darkness to achieve dark-adapted vision the more stars can be seen until, well away from light pollution, your view of the sparkling and twinkling jeweled infinity becomes breathtaking. Never mind that we cannot see all of the 100 billion stars in our galaxy, The Milky Way; just observe in awe the many hundreds of visible specks, each with its own structure, history, and story to tell.

Figure 10.2 is a remarkable image of the Milky Way over Spain, and over the clear skies of Portugal even more stars can be seen in this wonderful narrower field view (*see* Fig.10.3). It is likely that the most we will ever know about these other worlds will be from their brightness, movement, and interpretation of their spectra.

T. Buick, *The Rainbow Sky*, Patrick Moore's Practical Astronomy Series, DOI 10.1007/978-1-4419-1053-0_10, © Springer Science + Business Media, LLC 2009

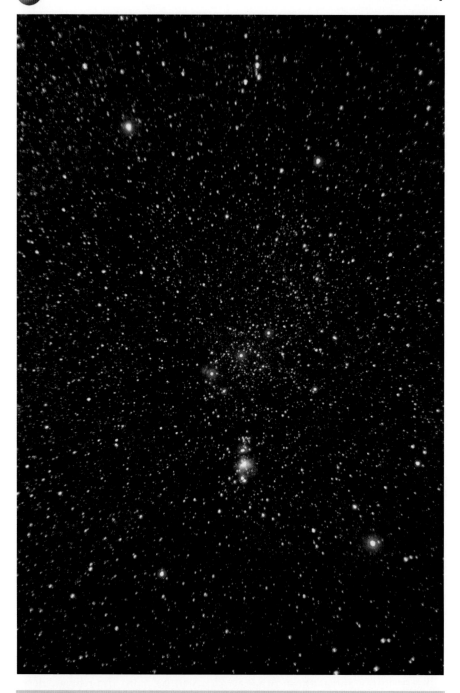

Fig.10.1. Winter sky over Dungeness, England (Credit: Mark Shelley, Orpington Astronomical Society).

Table 10.1.

Star	Astronomical name	Brightness	Note
Capella	Alpha Auriga	0.08	
Rigel	Beta Orionis	0.12	
Betelgeuse	Alpha Orionis	0.0–1.3	Red supergiant
Vega	Alpha Lyrae	0.03	Known as the
Altair	Alpha Aquilae	0.77	summer triangle
Deneb	Alpha Cygni	1.25	
Sirius	Alpha Canis Major	−1.46	Brightest star
Dubhe	Alpha Ursae Majoris	1.8	Brightest star in "The Plough"
Schedar	Alpha Cassiopeia	2.2	Brightest star in Cassiopeia
Castor	Alpha Geminorum	1.6	The Twins
Pollux	Beta Geminorum	1.14	
Alnilam	Epsilon Orionis	1.7	The stars of Orion's
Alnitak	Zeta Orionis	2.0	belt
Mintaka	Delta Orionis	1.9–2.1	

It is a mind-exercising experience to peer at a bright white spot in the sky and understand what it means in terms of the known complex physical laws – the massive size, the distance, its evolution, its color, and its implications for the structure of the universe – and our own very existence. But it is possible to have a bit of fun, too, with that apparently white spot!

Fig.10.2. Like smoke billowing from a chimney, the band of stars that is our Milky Way (Credit: Mark Shelley, Orpington Astronomical Society).

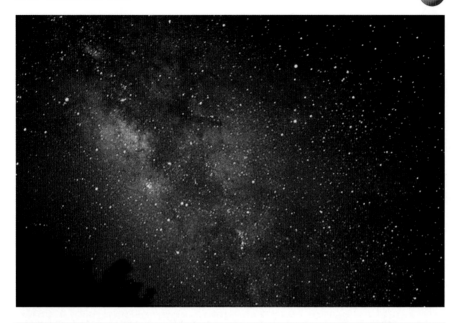

Fig.10.3. The Milky Way in dark skies over Portugal (Credit: Paul Whitmarsh, Orpington Astronomical Society).

Defocusing

The atmosphere can act as a refracting medium and on the macro scale sunsets produce great swathes of ever-changing blankets of color. At the smaller scale, light from the tiny points of stars is affected by turbulence in the air, owing to the existence of tiny "cells" that can be as small as a tenth of a centimeter. The cells have their own properties of temperature, refractive indices, speed, and so on that distort the incoming light to cause twinkling, more properly known as scintillation. Astronomical observers refer to the amount of scintillation or turbulence as the "seeing" condition. The less the turbulence the better the "seeing", and the better will be the sky pictures captured. The larger the visual diameter of a bright spot and the angle subtended at the eye, the less twinkling is observed. The near planets, therefore, twinkle less than stars.

Defocusing on a star can not only be a very thorough way of testing the performance of a telescope (*see* the Appendix of this book) but also help to make the color of that star more apparent. Figure 10.4 shows a matrix of several defocused stars. A variety of methods have been used to spread the light from the small bright spot to reveal actual star colors to the eye and brain. For example, David Malin arranged

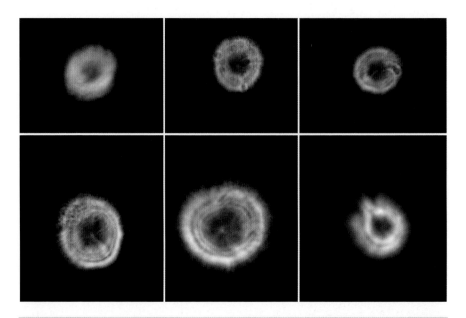

Fig. 10.4. A hint of color from defocused stars.

for a gradual, stepwise, automated defocus during a timed exposure to produce an exciting and artistic star trail presentation of the colors of Orion (again,*see* the Appendix of this book) and another such trail of the Southern Cross (*see* Fig.10.5).

Other simple methods can be employed to capture and display the scintillating colors. Here is one. First of all view Sirius or another bright star through your telescope and note the improved view of – and the rapidly changing – colors. The more atmosphere the light has to travel through the more turbulent cells the light must pass through. So, stars near the horizon will give the best chance of refraction.

As the twinkling is obviously a very fast process, a neat way of capturing what is going on is to append a digital camera to the telescope and take a movie of several seconds duration. Experiment with various eyepieces and camera zoom options to optimize the capture. Watching the camera back-screen is exciting enough, but upload the movie to a computer and play. Pause and move forward frame by frame to see the changing patterns of vivid colors.

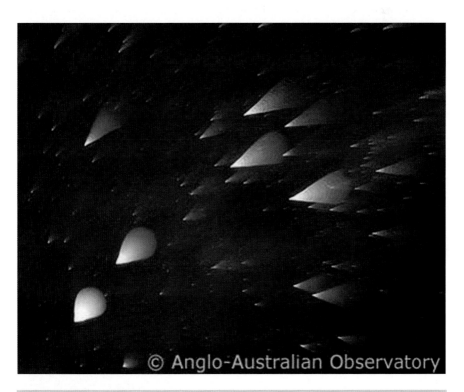

Fig.10.5. Auto-defocus trail of the Southern Cross (Credit: With permission from David Malin, Anglo Australian Observatory).

Using a movie-editing application, you can extract individual frames for a permanent record of proof of twinkling (*see* Fig.10.6). The colors will be characteristic of that star, but attempts to use these as a substitute for prism or diffraction grating observations are probably not useful.

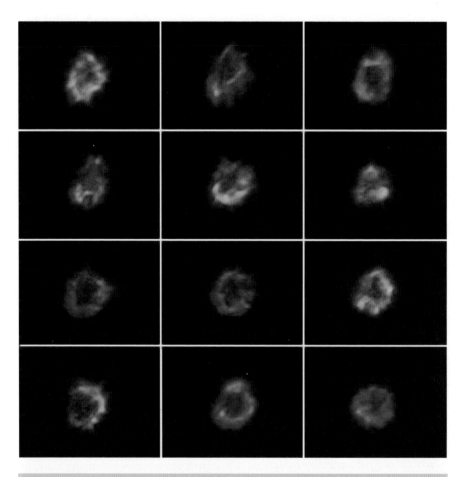

Fig.10.6. Still frames from the movie of Sirius.

Trails

Here is a more unusual astrophotography technique. Out on your patio or garden under a clear, dark night sky set up your digital camera on a tripod (no telescope required here). Film camera will do, although you will have the usual wait for development, assess, and re-do. Set to a zoom of about 10×, or maximum if less than this is available, and use a focus of infinity. Point the camera at Sirius (Alpha CMa in Canis Major), the brightest star in the sky and near the very recognizable constellation of Orion, or a similarly bright one in your own sky location and time.

Now give the tripod a very gentle kick while watching the viewing window! (This is the opposite of recommendations for star trails, where keeping still is paramount). Actually, a controlled but rapid nudge is safer! A glorious-colored thread is produced. Such photographs can be called "kick trails" and were first noticed by this author following an accidental trip in the dark! Take a photograph of a few seconds exposure, even 1 or 2 s will do with practice, to produce a good kick trail (*see* Fig.10.7).

Standardizing the "kick" helps you to create a (very) rough guide to comparison of star types, such as the blue of Orion's belt and the gold of Betelgeuse (*see* Fig.10.8). The two- and three-pointed kick trails shown in Fig.10.9 were achieved by lifting each leg of the tripod slightly during an exposure of a few seconds, the increased trail length allowing the eye to perceive color more clearly.

Some interesting kick trails can be captured quite unexpectedly. Figure 10.10 was obtained while repositioning the camera during 30-s exposures to catch the ISS speeding overhead. The number of different ways of producing kick trails are obviously limitless, and again for fun, this author built a contraption to enable a circular movement over a period of around 30 s (*see* Figs. 10.11 and 10.12), although not producing such pleasing graphics as for Sirius kick trails.

Figure 10.13 shows a double circle for Sirius. It might be much better to see the traditional rotations, "waiting for the sky to drift" during a long exposure, only satisfactorily captured under reasonably dark skies (*see* Fig.10.14).

It is worth recalling that even simple zooming in on 30-s exposure images on the computer afterwards often reveals the color, as shown in Fig.10.15, for the hot, white/blue, main Pleiades stars.

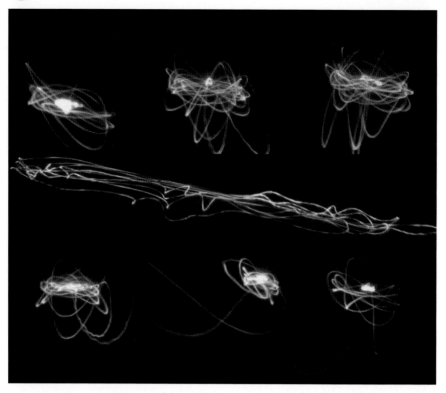

Fig.10.7. Some Sirius "kick trails."

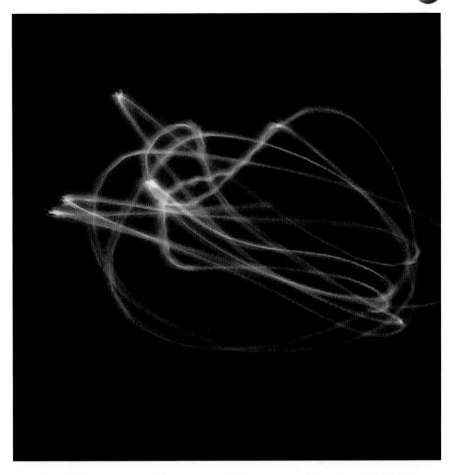

Fig.10.8. Betelgeuse gold "kick trail."

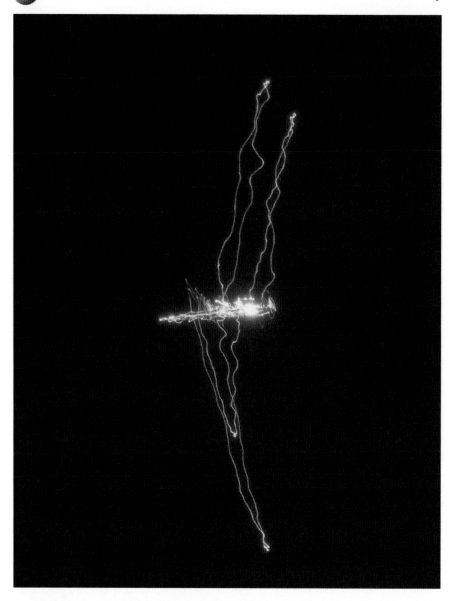

Fig.10.9. Betelgeuse golden "bi-trail."

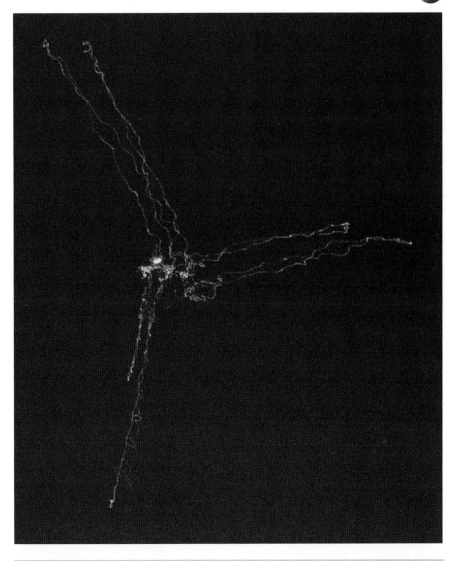

Fig. 10.9. (continued)

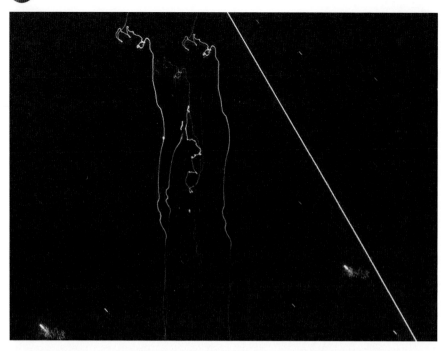

Fig. 10.10. Accidental trails during capture of an ISS satellite trail.

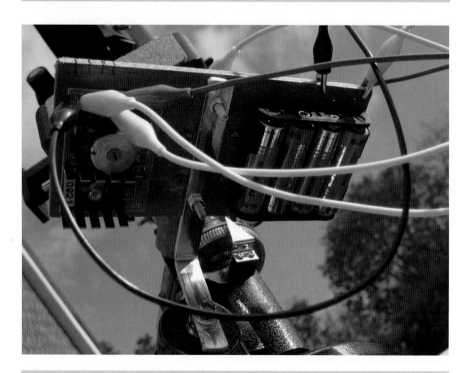

Fig. 10.11. Contraption for imaging star circles.

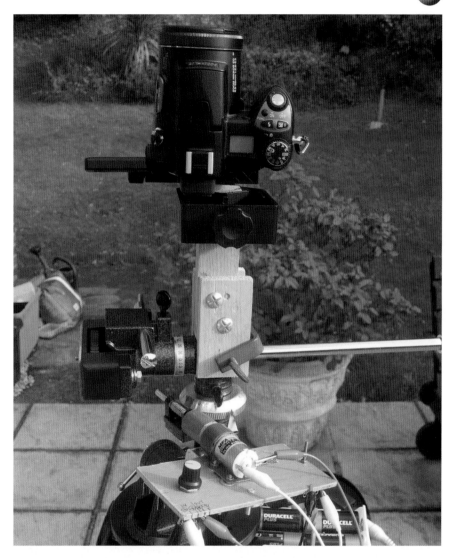

Fig. 10.12. Contraption for imaging star circles.

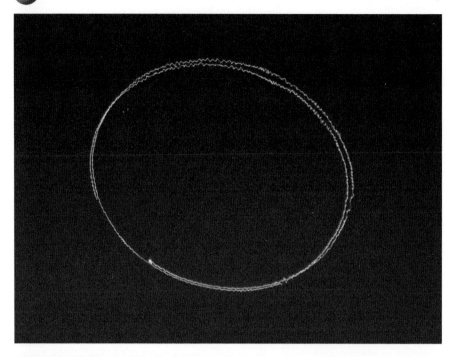

Fig.10.13. A double circle of Sirius.

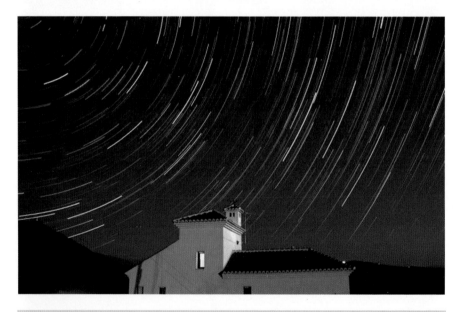

Fig.10.14. Time exposure to produce polar circles (Credit: Mark Shelley of the Orpington Astronomical Society).

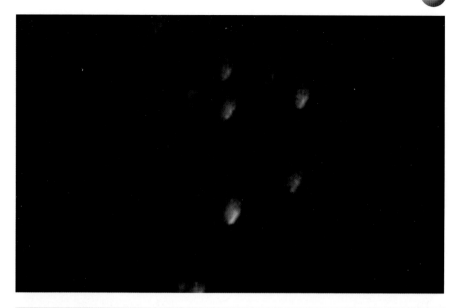

Fig.10.15. Just a short time exposure, 30 s, is sufficient "sky movement" to allow star colors to be revealed using an off-the-shelf compact camera.

CHAPTER ELEVEN

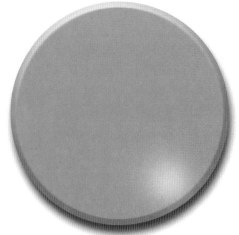

A Few Last Thoughts

So there it is! The universe is full of color – actually full of radiation of different types, as represented by the electromagnetic spectrum (*see* Fig. 11.1). As Einstein discovered, mass and energy can be converted one into the other ($E = mc^2$), so we may consider ourselves as composed of the energy of light or radiation!

The previous chapters have included many images to illustrate the existence and structure of the huge variety of solid and gaseous entities in the universe, as well as many phenomena from the ground level of Earth to the furthest reaches of telescopic vision. Since the speed of light is "so slow," the images also span time from the here and now to many billions of light years ago, not that long after the Big Bang. The range of images also provides a summary of what current technology is capable of, allowing us to see or know (Fig 11.2).

Development of that technology races on at such a pace that it is impossible to keep totally up to date with what is going on and being discovered. A telescope is now a tool that could observe the whole sky and at all wavelengths of radiation simultaneously, making it possible to view even closer to the time of the Big Bang. The Large Hadron Collider might reveal evidence for the Higgs boson or particles previously unknown or only guessed at, causing a flurry of research activity to come up with a completely new idea of the rules governing the origin and structure of

T. Buick, *The Rainbow Sky*, Patrick Moore's Practical Astronomy Series, DOI 10.1007/978-1-4419-1053-0_11, © Springer Science + Business Media, LLC 2010

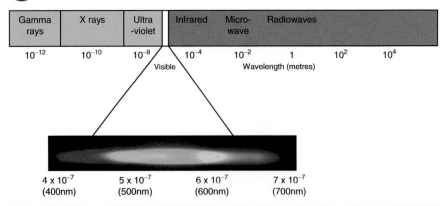

Fig. 11.1. The electromagnetic spectrum.

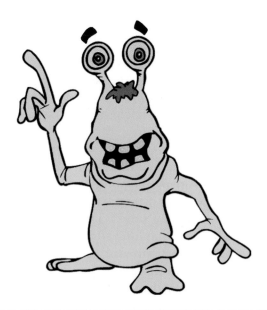

Fig. 11.2. Friendly or hungry?

the universe. Maybe an explanation of dark matter could pop out of the observations and equations. Maybe we will get to know enough about newly discovered planets to find an Earth look-alike. Will we try to make contact? The future is very exciting – or frightening (*see* Fig. 11.3).

Accessibility of the ever-increasing sophistication of off-the-shelf digital cameras, especially when used in conjunction with the ubiquitous

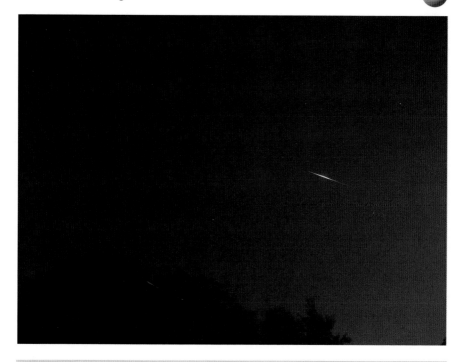

Fig. 11.3. The flare of iridium 23 against a Moonlit sky.

home computer, has made photographers of most of us. Millions of images available from the websites of many amateur photographers and astronomers as well as from the national and international agencies and institutions reveal the events and phenomena in our wonderful world and beyond. Without such windows on nature how many would have heard of fogbows, earthbows, anticrepuscular rays, volcanic lightning, Echus Chasm, Brockenspecter, glories, circumzenithal arcs, stereo space missions, ice volcanoes on Enceledus, Local Group galaxies, and images from almost as far back as the beginning of time?

To illustrate the accessibility of producing such images a photograph was taken with an off-the-shelf digital camera, without using zoom or shutter delay, to capture an iridium flare. The exposure was set to 15 s, which is available now on most cameras. The time of arrival of the satellite was found from an appropriate website, as was the direction and altitude. The necessary hands-off setup was accomplished by fixing the camera to a tripod, but it could easily have been wedged at an angle some other way. Just 5 s prior to the expected flare event the shutter was operated. A slight wobble at this stage would not affect the image, as the view was

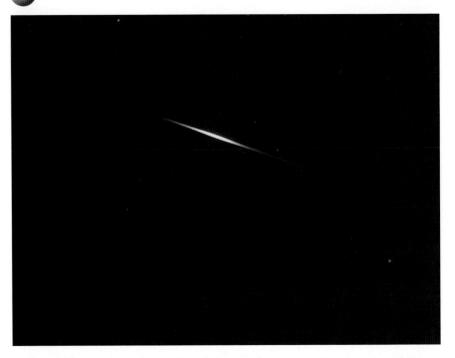

Fig. 11.4. Computer magnification to see more detail of the iridium 23 flare.

of a dark sky initially. The first image, Fig. 11.3, shows the flash with a background of Moonlit trees and the second a computer-cropped image to showcase the profile of the flare (Fig 11.4).

Wherever you are on this Earth something colorful will be happening in the day or night sky or around you in the atmosphere. All you have to do to witness such an event is to keep looking. But to prove to others that you have seen it, take a photograph for a record. There are millions of examples, and a sample of them is showcased in previous chapters. There are rainbows from Madeira to Hawaii, aurora from the Arctic to the Antarctic, Sun dogs, crepuscular rays, and lightning from Madagascar and New Zealand to England. Sunsets are everywhere. The more common the feature the more important it is to ensure the best example is captured and, if possible, include a meaningful or attractive landmark to frame it or add to the story in the picture. You might even be able to capture a rare occurrence such as an aurora in low(ish) latitudes (*see* Fig 11.5).

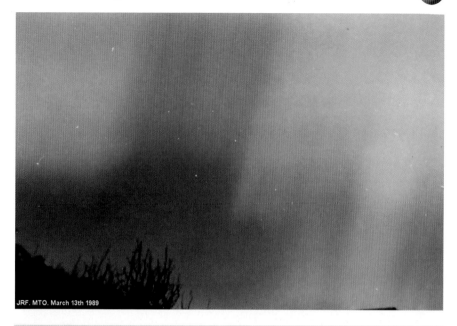

JRF. MTO. March 13th 1989

Fig. 11.5. A chance capture of an aurora in southern England (Credit: John Fletcher, Mount Tuffley Observatory).

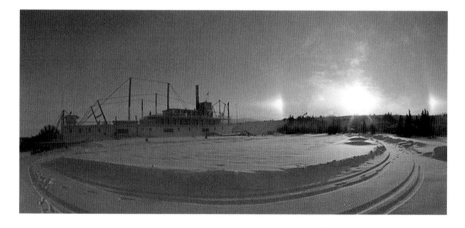

Fig. 11.6. Sun dogs over a historic boat wreck (Credit: David Cartier).

Two poignant images from the past particularly stand out as an example of what you can achieve: Sun dogs over a Yukon river boat wreck, Fig 11.6, and the color-slide photograph, Fig 11.7, from the mid-1950s of a sunset

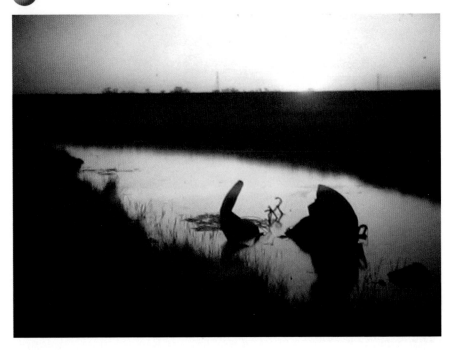

Fig. 11.7. A reminder of the bleakness of a very cold winter and the tragedies of human conflict.

picking out the casing of an exploded mine just beyond the beach. The coldness of the scene is supported by the fact that the sea at Seasalter in Kent, England, was frozen during that winter when the photograph was taken. David Cartier writes about the boat wreck sunset with parhelic arcs (Sun dogs) seen adjacent to the beached historic riverboat Klondike, along the Yukon river at Whitehorse, Yukon Territory, March 5, 2007. A happier reminder is to keep watch on the sky above perchance to witness the colors of a circumzenithal arc (*see* Fig 11.8).

It may be a little daunting for a novice to begin taking wonderful photographs or even seem impossible, but through the use and awareness of a few simple techniques and rules the impossible can be achieved. Of course, Hubble type images *are* impossible for most of us, but beautiful atmospheric phenomena and night sky photographs through amateur telescopes are well within the reach of any enthusiast. As the photographic capture of particular sightings becomes more ambitious the composition of the image is high on the agenda. Witness the double iridium flare in a

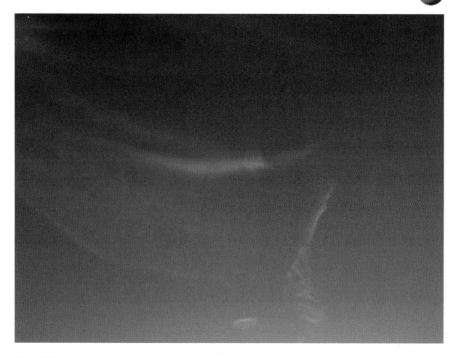

Fig. 11.8. A lonely strip of color directly above (Credit: David Fitzgerald).

previous chapter and the Hale–Bopp comet against an aurora backdrop (*see* Fig. 11.9).

Three-for-one images are also there to be captured as in the wonderful star trails on a canvas of zodiacal light and the Milky Way (*see* Fig 11.10), and how about a combination of a fogbow, glory, and Brockenspecter (*see* Fig. 11.11).

For the beginner how exciting it is to point out to a neighbor that the ISS will be traversing the overhead night sky in 1 min! And then take a photo of the trail! Magic! Is a colorful "kick trail" for the levity of the expert or amazement of the novice? Not sure, but it can be used to illustrate and introduce much of the physics of stars and the Earth's atmosphere.

Light, its detection and manifestations, has come a long way since early human's fires, although it has taken thousands of years. Landmarks in its development can be picked out. Ancient oil lamps were a step forward, then came gas with the beautiful gasoliers and electric domestic lighting. A brilliantly crafted triple color lamp was a great example from the era

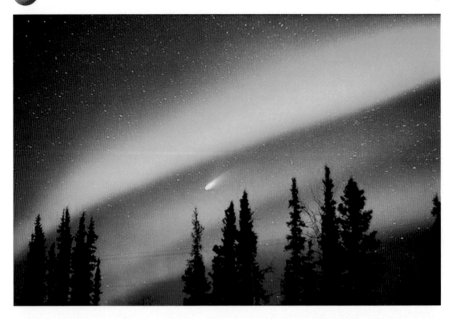

Fig. 11.9. Comet Hale–Bopp and aurora, Tok, Alaska (Credit: David Cartier).

Fig. 11.10. Star trails, zodiacal light, and the Milky Way (Credit: Joe Cali).

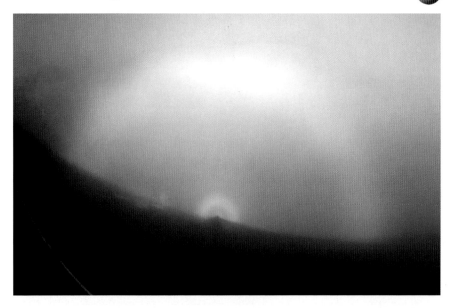

Fig. 11.11. Three in one: fog bow, glory, and Brockenspecter (Credit: Mila Zinkova, http://home.comcast.net/~milazinkova/Fogshadow.html).

of the development of color photography for permanent color records (*see* Fig.11.12).

Newton really did start something when he elucidated the composition of white light. Just the seven colors were enough to get everyone thinking and led to knowledge of the massive range of radiation from gamma rays to radio waves. Although an understanding of the origin and perception of color helps us to appreciate the natural laws behind a rainbow, halo, and glory, sometimes just sit back, relax with a friend, soak up the wonder of a sunset wherever you are in the world, and say WOW!

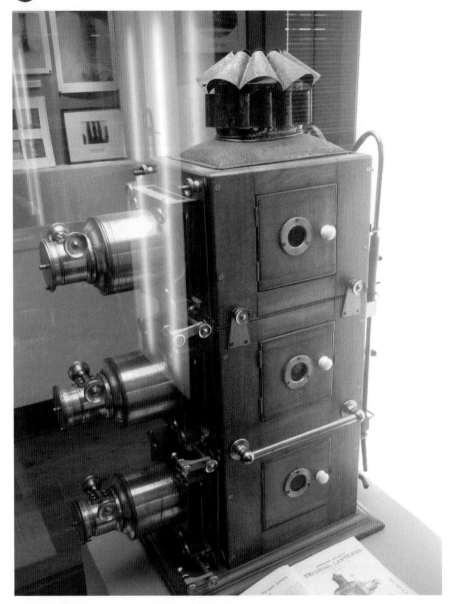

Fig. 11.12 A triple color projector (Credit: Birr Castle, County Offaly, Ireland).

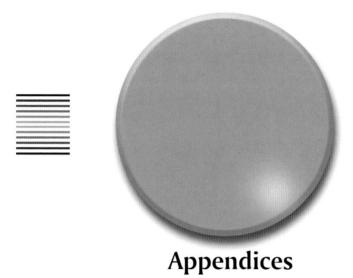

Appendices

Appendix 1: References

General

Includes publications containing information for many of the topics covered.

Minnaert, M.G.J. *Light and Color in the Outdoors,* Springer, New York, 2004. Description of solar nuclear reactions and other properties of the Sun.

Lang, Kenneth R. *The Cambridge Encyclopedia of the Sun,* Cambridge University Press, London, 2001.

Moore, Patrick. *The Data Book of Astronomy,* Institute of Physics Publishing Ltd., London, 2000.

Moore, Sir Patrick. *Atlas of the Universe,* Philips, 2007.

May, Brian, Patrick Moore, and Chris Lintott. *Bang, The Complete History of the Universe,* Carlton Books Ltd., London, 2006.

Ridpath, Ian. *A Dictionary of Astronomy,* Oxford University Press, London, 1997.

Chapter 1

http://www.photographywebsite.co.uk. Helpful advice on digital photography:

http://www.digital-photography-school.com/blog/how-to-photograph-sunrises-and-sunsets/. Tips for photographing sunsets:

Chapter 2

http://www.astrophysicsspectator.com/topics/stars/FusionHelium.html. About helium fusion.

http://www-istp.gsfc.nasa.gov/stargaze/Q8.html. Quantum tunneling effect:

Chapter 3

http://www.planetquest.jpl.nasa.gov. For accurate measurement of cosmic bodies in a search for terrestrial planets, SIM Planet Quest.

Robinson, Keith, *Spectroscopy: The Key to the Stars,* Springer, New York, 2007. An introduction to light and atomic structure.

Chapter 4

http://www.vendian.org/mncharity/dir3/starcolor/details.html. An article on the color of stars by Mitchell Charity.

http://www.en.wikipedia.org/wiki/Stellar classification. On star classification.

Chapter 5

Tonkin, Stephen F. *Practical Amateur Spectroscopy:* Springer, London, 2002. A description of diffraction:
http://astro.u-strasbg.fr/~koppen/spectro/spectroe.html
jerryzhu@gmail.com
Details of how to obtain excellent photographs of spectra using the "cereal box" technique can be found on these websites by Joachim Köppen and Jerry Xiaojin Zhu.

Chapter 6

http://www.nikondigital.org/articles/rgb_digital_camera_color.html. For an introduction to understanding digital cameras and how they reproduce color.
http://www.sdss.org. The Sloan Digital Sky Survey.
Brecher, Kenneth. "Galaxy" World Book Online Reference Center, 2005, World Book, Inc. http://www.worldbookonline.com/wb/Article?id=ar215080. For a full description of galaxies.
http://www.galaxyzoo.org. The Galaxy Zoo project.
http://www.birrcastle.com. The Birr Castle website for the "Leviathan of Parsonstown" telescope and the historic science center.
Herbst, Eric. Composition of Interstellar Gas. Chem. Soc. Rev. 2001, **30**, 168–176.
http://www.umbra.nascom.nasa.gov/solar missions.html. About solar missions.

Chapter 7

Apt, Jay, Michael Helfert, Justin Wilkinson, Ed., Roger Ressmeyer, "NASA Astronauts Photograph the Earth: *ORBIT,*" National Geographic Society, 1996.
Buick, Tony. *How to Photograph the Moon and Planets with Your Digital Camera,* Springer, New York, 2006.
http://www.atalaia.org/filipe/moon/colorofthemoon.html. Details and instructions on image saturation techniques from Filipe Alves.
http://www.esa.int/esa-mmg/mmg.pl?b=b&type=I&mission=SMART-1&single=y&start=68. SMART-1 mission to the Moon:

Chapter 8

There are many websites for anyone to post and view images. An example is http://www.flickr.com

http://www.cloudappreciationsociety.org. The Cloud Appreciation Society:

http://www.en.wikipedia.org/wiki/. Green Explanation of the Sun's green flash.

http://www.philiplaven.com/p2c1a.html. How glories are formed.

Chapter 9

http://www.heavens-above.com. For prediction of arrival of ISS, Iridium, and other satellites:

Russell, Michael S. *The Chemistry of Fireworks*, Royal Society of Chemistry, Cambridge, England, 2000.

Chapter 10

http://www.willbell.com/tm/tm5.html. Telescope performance testing through defocusing.

Appendix 2: Glossary

The explanation for the many cloud types and commonly understood phenomena are not listed here. Information can be found via the references given or the Internet.

Absorption spectrum. A spectrum that has dark lines or gaps. The absorption spectrum of the elements in a gas cloud in space, for example, will be seen when a beam of light from a hot source behind shines through the cloud and some photons are absorbed at characteristic wavelengths. So the position of the lines reveals the identity of the elements.

Albedo. The surface reflectivity of a nonluminous body or the fraction of incident light or radiation that is reflected. Although the Moon can appear to be almost blindingly bright at full Moon, it is a poor reflector with an albedo of just 7.9%. This is an average value, since the Mare, or "seas", are darker than the highland areas. Again, on average, Earth's albedo is around 30% although surface and cloud feature variations have a large effect. Fresh snow can have a value of 95%, while soil and coniferous forests can be as low as 0.5%. Note that albedo is not exactly the same as reflectivity because it takes into account reflection from an uneven or granular surface and is actually known as diffuse reflectivity.

Alexander Band. Alexander of Aphrodisias, ~ A.D. 200, was possibly a student of Aristotle in ancient Greece. He is commonly recognized as the first to describe the dark band between a primary and secondary rainbow. Light that is deviated between the angles leading to the two bows can never reach the observer and so appears as a darker area.

Aphelion. The position on an elliptical orbit that is furthest from the Sun's center.

Apogee. The position on an elliptical orbit that is furthest from the Earth's center.

Argand burner. Argand lamps or burners have a fuel reservoir above and to the side of the burner. The burner is fed fuel horizontally by gravity through an arm or arms that extend between the font (reservoir) and the burner. The device was invented by Ami Argand of Switzerland, whose oil lamp featured a burner with a tubular wick that allowed air to flow up through the hollow middle of the wick as well as around the outside, making oil flames burn brighter, possibly ten times brighter, than any other existing technologies of the day.

Asterism. A recognizable pattern of stars that forms part of one or more constellations of stars such as the Plow or Big Dipper within the constellation of Ursa Major.

Atom. The smallest component of an element that retains the properties of that element. An atom is a fundamental component of matter.

Astronomical Unit (AU). Approximates to the mean distance between the Earth and the Sun. The modern, more technical definition puts the average distance from Sun to Earth at 1.000,000,031 AU.

Aurora. An aurora is an effect seen in the atmosphere when charged particles from the radiation belts surrounding the Earth are accelerated down the magnetic field lines near the North and South poles. They collide with atoms of gas in the atmosphere such as nitrogen and oxygen, causing them to emit light at characteristic wavelengths. Low gas densities are required, and hence many eerie colors are seen at 70–300 km above the ground.

Baryonic. "Ordinary" matter, that is, matter composed of atoms. Until a few decades ago, it was thought that the entire universe was composed of baryons, but evidence now suggests that there is something or things that we cannot see, often referred to as dark matter.

Bioluminescence. The light produced by a chemical reaction within an organism. At least two chemicals are required. The one producing the light is known as luciferin, and the other that catalyzes the reaction, luciferase.

Black Hole. A theoretical (although there is much evidence for it) entity predicted by the equations of general relativity; it is formed when a star of sufficient mass undergoes gravitational collapse. The highly compressed mass occupies a small area, causing a singularity or infinite space–time curvature that allows nothing to escape its event horizon, not even light. Black holes come in many sizes, from supermassive, equivalent to thousands of times the mass of the Sun, to tiny ones produced as energetic cosmic rays hit the atmosphere. It is thought that a supermassive black hole exists at the center of our galaxy.

Caldwell catalog. A catalog of 109 bright star clusters, nebulae, and galaxies in the sky created by Sir Patrick Caldwell-Moore to help amateur astronomers and first published in the magazine *Sky & Telescope* in December 1995.

Color. Electromagnetic (EM) radiation. Visible color is in the small part of the EM spectrum that spans wavelengths from around 400–700 nm.

Color index. The difference in a star's magnitude at two different wavelengths, such as viewed through blue and visual color filters. Hence the B–V color index.

Constellation. The celestial sphere, the sky, is divided into 88 areas called constellations. Each constellation forms a recognizable pattern of stars to which is ascribed the name of a mythological or other figure. Often a particular part of the constellation is more easily recognized and is known as an asterism.

Crepuscular. Relating to or referring to twilight, the time between day and night at dawn or sunset.

Danjon scale. A five-point scale of darkness to help define the approximate brightness, often redness, of the Moon during a lunar eclipse. The score is largely determined by the properties of the atmosphere at the time. Score "0" = darkest, "4" = brightest.

Dark matter. A certain amount of matter is required to accommodate the suggestion that our universe is finely balanced between being open and closed, although opinions vary on how much this is. The total amount of baryonic matter can be estimated but only accounts for a few percent of what is required. The remainder is known as dark matter and dark energy, and although its composition is unknown, there are many indications of its existence, such as the behavior or movement of galaxies.

Deuteron. The nucleus of a deuterium atom, an isotope of hydrogen, that contains a neutron bound to a proton.

Diffraction. The result when a wave encounters an obstacle. Although this refers to any wave, it is light wave that is referred to in the text and the generation of color in particular. Diffraction is the slight bending as light passes the edge of objects such as water droplets or the interface between two media with different refractive indices. Optical effects resulting from diffraction are produced as a result of the interference of light waves.

Doppler effect. The change in wavelength of electromagnetic radiation with the change in relative motion between the source and the receiver. If the source is moving away from the observer then the wavelength is lengthened or red-shifted. If moving toward, the wavelength is shortened or blue-shifted.

Electromagnetic (EM) spectrum. The complete range of electromagnetic radiation that includes radio waves, which are longest, to the highly energetic short wavelength gamma radiation. Around the middle, from about 400 to 700 nm, occurs the visible spectrum.

Electron. An elementary particle that hosts a negative charge. Electrons orbit the nuclei of atoms in energy levels, and the number of electrons balances the number of positively charged protons in a neutral atom.

Element. A substance that comprises atoms of the same type. There are 92 naturally occurring elements.

Emission spectrum. The relative intensity of electromagnetic radiation of each wavelength a body emits when excited, often due to being heated or irradiated. The pattern of emissions, or brightly colored lines, is characteristic of each element and is, therefore, useful in identifying the composition of a star.

Fraunhofer lines. The dark lines or bands in an absorption spectrum characteristic of the elements causing them. First discovered by Joseph von Fraunhofer in 1814.

Galaxy. A group or system of stars that are gravitationally bound that also include gas, dust, and, probably dark matter. They range from dwarfs with fewer than a million stars to supergiants with billions of stars. They are usually classified according to their shapes, which include spirals, elliptical, and subcategories of these two.

Gamma rays. The shortest wavelength electromagnetic radiation. They are the highest energy photons. The most massive explosions in the universe may well be the origin of gamma ray bursts that are thought to be due to supernovae, the final scenario in the death of huge stars.

Gasolier. An ornate gas light made in the nineteenth century of brass or other metal. It looks like a chandelier with branches holding burners stemming from a central shaft that is hollow to allow gas to be piped through. A good example can cost tens of thousands of dollars.

Gliese catalog. Also known as CNS, this is a star catalog originally attempting to list all stars within 20 parsecs of Earth. CNS3, the three referring to the edition, lists all those within 25 parsecs and contains 3,803 stars.

Glory. A multicolored ring surrounding the shadow of the shape of a point on a cloud or other medium of uniformly sized water droplets. So, it could appear as a colored circle around the viewer's head or the shape of an airplane. It is produced through a combination of diffraction, reflection, and refraction.

Gravitational lensing. An effect of bending light around an object due to gravity. The huge gravity generated by massive galaxies allows multiple images to be seen of a bright object behind the galaxy. From known parameters of the galaxy, it is possible to estimate the distance to the bright object behind, such as a quasar, which emits enormous amounts of energy.

Higgs boson. The Higgs boson field is a mechanism suggested by the English physicist Peter Higgs in the 1960s to explain how particles acquire the properties associated with mass. The Higgs boson itself has not yet been detected, but it is hoped that the huge energies needed to find it or infer its existence will be provided by the Large Hadron Collider.

H–R diagram. The Hertzsprung–Russell, or H–R, diagram plots luminosity against temperature (or class or spectral type) for stars. The majority of stars fall on a curving diagonal line called the main sequence. As stars live out their lives, changes in the structure of the star are reflected in changes in stars temperatures, sizes, and luminosities that cause them to move along the plot of the H–R diagram.

Incandescent. Light is produced by incandescence when it comes from a heated solid. For example, when a bar of iron is heated, it first becomes red, then white. The process of turning such heat energy into light is called incandescence.

Index catalog. A catalog of galaxies, nebulae, and star clusters that supplements the NGC (New General Catalog).

Interference. An effect in which two or more sets of electromagnetic waves combine to either reinforce or amplify their total amplitude (constructive interference) or decrease it (destructive interference). This definition is pertinent to this work but can also have other meanings, such as unwanted radio signals.

Interstellar medium. The matter that fills the space between the stars. Although space is a very effective vacuum compared to any achievable on Earth, it still contains huge quantities of gas, dust, magnetic fields, and charged particles. This constitutes around 15% of visible matter in the Milky Way. The clouds of molecular and atomic hydrogen represent the raw material from which stars can be formed.

Iridescence. An optical phenomenon in which the color changes with the angle from which a surface is viewed. Butterfly wings and duck feathers are examples. Constructive interference is the basis for this effect.

Isotopes. Atoms that have the same number of protons, hence the same chemical properties, but different numbers of neutrons. Isotopes have different physical properties, although the differences may be small. Deuterium (with one neutron) is a slightly heavier isotope of hydrogen, and the tritium isotope (with two neutrons) is also radioactive.

Kelvin temperature scale. A thermodynamic scale of temperature adopted as the international standard. A degree Kelvin is the same as a degree Celsius, but the scale starts at absolute zero, the lowest temperature attainable, i.e., $-273.15°C$.

Kuiper Belt. An outer region of the Solar System from about 30 to 50 AU that contains a huge number of small "icy" bodies composed largely of water, ammonia, or methane "ice" and larger bodies up to the size of dwarf planets. Some of the best known dwarf planets are Pluto, Eris, Sedna, and Quaoar.

Lagrangian point. A point at which a small body can remain in the plane of two much larger bodies. Five such points are known, three of which lie in a straight line through the two large bodies, the Sun and the Moon. The other two points each form an equilateral triangle with the two large bodies and lie on the orbit of the smaller one, 60° in front and behind.

Large Hadron Collider. A high energy particle accelerator – the largest in the world – that is designed to accelerate protons and other particles to 99.999999% of the speed of light. The particles will be smashed together in an attempt to learn more about high energy physics with the hope of detecting indirectly the existence of the Higgs Boson that is hypothesized to be the cause of particles having mass.

Laser. Light amplification by stimulated emission. A laser light emits a coherent (in phase) low divergent beam of light.

Lenticular. Lens-like (i.e., galaxies or clouds) but can also mean pertaining to freckles.

Luminosity. The amount of radiation a star emits, taking into account interstellar absorption.

Lychnology. A study of ancient lamps and lighting.

Lychnoscope. A low side window in a church, often with a view of the altar, to allow a view inside the church to those such as lepers, who are not permitted to enter.

Magnitude. A measure of the brightness of a star. The smaller or more negative the number, the brighter the star.

Messier catalog. Originally a list of 103 sky objects that could be mistaken for comets compiled by Charles Joseph Messier. Each object has an "M" number, such as M42, the Orion Nebula.

Meteor, meteoroid, meteorite. Often confused, a meteoroid is a small particle in space from a comet or an asteroid leading to a meteor, which is a trail of light seen in the sky as the meteoroid enters and burns up in the atmosphere. A meteorite is a fragment of a natural object from space that is large enough to actually reach the surface of Earth without burning up.

Milky Way. The name of our own galaxy and also the band of stars seen across the sky, the appearance of which is caused by the dense mass of stars in the plane of our galaxy.

Nebula. A mass of gas in space that is often seen or illustrated as fascinating colors and shapes. Three types are described, "emission," "reflection," or "dark," referring to the origin or absence of light.

Neutrino. An elementary particle that has a minute mass, zero charge, and travels close to the speed of light. Due to these properties neutrinos can travel through matter with very little interaction and are, therefore, extremely hard to detect. Trillions of neutrinos pass through our bodies every second.

Neutron. A neutral elementary particle that occurs in the nucleus of all elements except the normal or lightest isotope of hydrogen. Outside a nucleus they decay into other elementary particles.

NGC. The New General Catalog of nebulae and star groups or clusters. The Andromeda Galaxy is NGC 224.

Nuclear fission. The splitting of a high atomic number element such as uranium to form elements with a lower atomic number, accompanied by the release of huge quantities of energy.

Nuclear fusion. The fusion of low atomic number elements to form an element of a higher atomic number, such as hydrogen to helium, with the release of energy.

Nucleus. The core of an atom, which consists of protons and neutrons, except the lightest isotope of hydrogen, which contains no neutrons.

Oort Cloud. A huge spherical outer cloud and disk-shaped inner cloud of cometary bodies surrounding the Sun. It is outside the Kuiper Belt and extends out to possibly 100,000 AU. This is most likely the source of all long-period comets such as Halley's comet.

Parsec. A unit of distance or length used in astronomy. It is the distance at which a length of one AU will subtend an angle of one arcsecond. The word is based on the parallax method of measurement, PARallax of one arcSECond. It is equivalent to 3.2616 light years.

Perihelion. The position in an elliptical orbit around the Sun that is closest to the Sun's center.

Perigee. The position in an elliptical orbit around the Earth that is closest to Earth's center.

Periodic table. An arrangement of all known elements in order of their atomic numbers. Elements show trends in their properties when viewed along rows or down columns. From the position of an element in the table clues as to properties of that element can be obtained.

Pharos. An ancient lighthouse built in the third century B.C. by Pharos of Alexandria.

Planck time. One of three fundamental physical constants: the speed of light, Planck's constant, and the gravitational constant. It is the time taken for light to travel one Planck length, or 16.163×10^{-36} m, i.e., 10^{-43} s.

Positron. The same as an electron but with a positive charge. It is the antimatter partner of the electron.

Proton. An elementary particle with a positive charge that occurs in the nucleus of all elements. The number of protons in an element defines its chemical properties.

Protanopia. One of the two types of red–green color blindness.

Purkinje effect. The tendency for the eye to perceive colors differently when comparing viewing in bright light with that at low levels of illumination.

Quantum theory. A theory in which energy levels can only exist in discrete quanta or packets. Quantum theory is the basis of modern physics and describes the behavior of matter and energy at the subatomic level.

Quarks. The theoretical building blocks of fundamental particles. The proton is composed of three quarks, two "up" and one "down." Quarks are not found on their own in nature.

Quasar. Highly energetic points of light that looked like stars, hence the name QUASi stellar, or quasar. They are the most distant objects known and are probably active galactic nuclei powered by supermassive black holes.

Radiation. Energy being transmitted as electromagnetic waves, hence X-radiation or microwave radiation.

Rayleigh scattering. The scattering of light by particles smaller than the wavelength of light (electromagnetic radiation). The extent of scattering increases as the wavelength gets shorter. Therefore, blue light is scattered more than red light and accounts for the blue color of the sky.

Roche limit. The orbital distance from a planet within which a satellite held together only by gravity, no tensile strength, will be torn apart by tidal forces. So man-made satellites can operate well within the Roche limit.

Rosetta Stone. An ancient stone, from 196 B.C., bearing the same passage of writing in three different Egyptian and Greek languages. Comparing translation attempts of each language helped decipher the others, contributing greatly to the translation of hieroglyphic writing.

Scintillation. A rapid and continuous change in the brightness of a star due to the effects of the light passing through a turbulent atmosphere. Also, the emission of flashes of light following bombardment by high energy particles of certain materials.

Seeing. The astronomical term for how well stars and other night sky objects can be seen. It depends very much on the properties of the atmosphere at the time as well as the extent of light pollution.

Sidereal. Relating to stars or measurements relating to stars.

Small field tritanopia. The inability to detect color due to the small angle of the light subtended at the eye.

Spectroscopy. The technique of obtaining and studying spectra. Especially useful for astronomers to determine the chemical composition and movement of stars.

Spectrum. The range of electromagnetic energies displayed in order of their wavelengths. Hence, the electromagnetic spectrum.

Star class. Stars are classified by their spectra and their temperature. There are seven main classes – designated O, B, A, F, G, K, M – for which a well-known mnemonic is "Oh be a fine girl, kiss me"!

Sun dog. A colored patch in the sky resulting from refraction and reflection of sunlight from cirrus clouds. The proper name for a Sun dog is a parhelion.

Supernova. A violent explosion resulting from the "death" of certain stars such as massive ones or white dwarfs in a binary system.

Synodic period. The time interval between successive oppositions of a superior planet or conjunctions of an inferior one.

White dwarf. A small, dense body, resulting from the death of most stars. Its density is equivalent to packing the Sun into a volume the size of the Earth.

WIMPS. Weakly interacting massive particles. These are hypothetical particles that are postulated to be the stuff of dark matter.

Wolf-Rayet (WR) stars. Very hot, large stars with strong emission lines in their spectra.

Zodiacal light. A faint glow in the sky seen an hour or so after sunset or before sunrise. It is probably a reflection from dust in the plane of the Solar System.

Appendix 3: Scientists and philosophers

Here are brief biographies of some of the people mentioned in this text.

Airy, George Biddell, 1801–1892. Seventh Astronomer Royal. Well known for his meticulous treatment of positional astronomy and installation of the transit telescope at Greenwich Observatory.

Abd al-Rahman al-Sufi (a.d. 903–986). A Persian astronomer who published a *Book of Fixed Stars* that included their magnitudes, positions, and colors. He recorded the earliest observation of the Andromeda Galaxy, a "small cloud."

Aristotle (384–322 b.c.) An influential Greek philosopher who wrote about many subjects, physical sciences in particular, and described a geocentric model of the universe.

Bacon, Roger (1220–1292). His date of birth is somewhat uncertain. Possibly the earliest founder of modern, empirical-based European philosophy and lectured on Aristotle at Oxford University in England. He recognized the visible spectrum centuries before Newton and his prisms. He said that only experience teaches anything.

Bohr, Niels Henrik David (1885–1962). An influential physicist who made a significant contribution to the theory of quantum mechanics.

Bok, Bartholomeus Jan (1906–1983). A Dutch–American astronomer who studied the structure of the Milky Way and the radio-wavelength mapping of our galaxy. He first described Bok globules, dark, dense clouds of dust and gas in space.

Brand, Hennig (c. 1630–1710). A German merchant and alchemist who searched for the Philosopher's Stone to turn base metals into gold, as did other alchemists at that time. After leaving service in the army he pursued alchemy and, in a wonderful display of fire and light from heating boiled-dry urine, discovered phosphorus (light bearer).

Boyle, Robert (1627–1691). An Irish scientist best known for his formulation of Boyle's Law and possibly the founder of modern chemistry.

Bunsen, Robert Wilhelm Eberhard (1811–1899). A German chemist most remembered for his design of an improved laboratory burner. Much of his work was devoted to photochemistry and emission spectroscopy.

Cannon, Annie Jump (1863–1941). An American astronomer who began work under E.C. Pickering to develop a classification scheme for stars. She was the leading contributor to the development of the Harvard Draper spectral catalog.

Cassini, Giovanni Domenico (1625–1712). An Italian-born French astronomer who became the first director of the Paris Observatory. He studied the moons of Jupiter, four of which he discovered. He discovered a large gap in Saturn's rings that is now named after him.

Champollion, Jean-Francois (1790–1832). With the help of the fundamental studies of others he deciphered Egyptian hieroglyphs, in particular part of the Rosetta Stone.

Déscartes, René (1596–1650). A French philosopher, mathematician, and scientist noted for, among other things, development of the Cartesian coordinate system of mathematics.

Doppler, Christian Johann (1803–1853). There is confusion as to Doppler's correct name. Was his middle name Johann or Andreas? According to his baptism certificate he was Andreas, but Johann is written on his gravesite in Venice. He is famous for the Doppler effect, a variation of electromagnetic frequency (sound in particular) between that received when an emitting body approaches compared with that when it is stationary or moves away.

Dreyer, Johan Ludvig Emil (1852–1926). A Danish astronomer who worked in Ireland initially as assistant to Lord Rosse, whose father built the huge 72-in Leviathan reflecting telescope. He is mainly noted for his creation of the New General Catalog of nebulae and star clusters.

Edison, Thomas Alva (1847–1931). An American inventor who invented an amazing number of things but is particularly noted for his development of a durable and practical light bulb and the principles of mass production.

Einstein, Albert (1879–1955). A German-born physicist most noted for his theory of relativity and what must be the most used and quoted formula of all time, E=mc², although he made many more contributions to science.

Euclid (320–275 b.c.). Also known as Euclid of Alexandria. Although dates of birth and death are quoted, they are not really known with any certainty. His most noted work was the development of Euclidean geometry.

Faraday, Michael (1791–1867). An English scientist or natural philosopher especially recognized for his work on electricity, electromagnetism, and electrochemistry.

Flamsteed, John (1646–1719). An English astronomer and appointed as the first Astronomer Royal or the King's Astronomical Observator. He resided at the Royal Observatory, Greenwich, England, for much of his life.

Fleming, Williamina Paton (1857–1911). She began her contribution to astronomy by assisting Edward Pickering (initially appointed as his maid) to classify the stars, the work being published as *The Draper Catalog of Stellar Spectra*.

Franklin, Benjamin (1706–1790). An American of many talents and especially active in the field of politics and international diplomacy. His scientific research and experiments with electricity are well known, particularly from the story of flying a kite into clouds to generate sparks. It is believed that he insulated himself at that moment to avoid electrocution, but he may not have actually ever carried out this experiment.

Fraunhofer, Joseph von (1787–1826). A German optician and physicist who discovered dark lines in the absorption spectrum of the Sun and made the first spectrometer.

Galileo Galilei (1564–1642). An Italian physicist, astronomer, philosopher, and mathematician. Best known for his support of the Copernican heliocentric model of our Solar System and his first use of a telescope to view the sky, the Moon, and Jupiter in particular. The four largest moons of Jupiter are named the Galilean moons in his honor. See the entry for Thomas Harriot below.

Gregory, James (1638–1675). A Scottish astronomer who first described a practical reflecting telescope containing two mirrors to get around the problem of chromatic aberration. He passed sunlight through a bird's feather to discover the diffraction grating and observed the splitting of sunlight into its component colors a year after Newton had achieved that with a prism.

Halley, Edmond (1656–1742). An English astronomer best known for predicting the periodicity of a comet and that it would return in 1758. It did and is now named Halley's Comet. He also predicted that the distance of Earth from the Sun could be measured from observations of the transit of Venus, which happened, but long after his death.

Harriot, Thomas (1560–1621). An English astronomer who is likely to have viewed the Moon through a telescope just before Galileo. Although he recorded his results at the same time as his friend the Welsh astronomer Sir William Lower, he did not publish them.

Heraclitus (535–475 b.c.). A Greek philosopher known for his teachings of change being central to the universe.

Herschel, John Frederick William (1792–1871). An English astronomer who mapped the southern skies with precision from Cape Colony in South Africa. The son of William Herschel.

Herschel, William (1738–1822). German born but traveled to England when he was 18 years old. Sometimes referred to as the founder of modern stellar astronomy, he discovered the planet Uranus and developed a theory of nebulae and the evolution of stars. He demonstrated that the Solar System moves through space.

Herschel, Caroline Lucretia (1750–1848). Sister of William and the first woman widely recognized as an astronomer.

Hertzsprung, Ejnar (1873–1967). A Danish scientist and astronomer. He discovered, in 1905, that the variations in widths of stellar spectral lines reveal that giant stars are of much lower density than main sequence or dwarf stars. His work led to the development with Henry Russell of the Hertzsprung–Russell diagram, first published in 1913.

Hipparchus of Nicea (Second Century b.c., possibly 190–120 b.c.) A Greek astronomer, philosopher, and mathematician born in what is now known as Turkey. Although much of his work has been lost he is credited with the founding of trigonometry and, being a dedicated and accurate observer, created possibly the first catalog of stars.

Hodierna, Giovanni Battista (1597–1660). A Sicilian astronomer, mathematician, scientist, and philosopher who practiced and developed much of Galileo's work. Most of his interest was in the Solar System bodies. He studied light through prisms and formulated his own description of a rainbow.

Hubble, Edwin Powell (1889–1953). An American astronomer who was the first to show that the universe is expanding. He also studied mathematics and law. He excelled at sports such as boxing and basketball and held the Illinois State high jump record. He made great contributions to the knowledge of galaxies and nebulae. The Hubble Space telescope is named after him.

Huggins, William (1824–1910). An English astronomer who studied the spectroscopy and chemistry of stars, galaxies, and nebulae.

Huygens, Christiaan (1629–1695). A Dutch physicist, mathematician, and astronomer who contributed much knowledge concerning the wave nature of light, accurate clocks, telescopes, and the rings and large moon (Titan) of Saturn.

Keenan, Philip Childs (1908–2000). With William Morgan and Edith Kellman, further developed the Harvard Classification of the Stars and known as the MK or MKK or Yerkes spectral classification.

Kirchhoff, Gustav Robert (1824–1887). A German physicist best known for his three laws of spectroscopy and, with his friend Robert Bunsen, established the principles of spectral analysis. He also contributed to the understanding of electrical circuits and the radiation emitted by heated objects.

Lacaille (1713–1762). A French astronomer (full name Abbé Nicolas Louis de Lacaille) noted for his catalog of around 10,000 southern stars and many nebulous objects observed from South Africa.

Lagrange, Joseph Louis de (1736–1813). Italian-born French mathematician most noted for calculations of the "three body problem" leading to the points of equilibrium related to the Sun, Moon, and Earth now known as Lagrangian points.

Legentil (1725–1792). A French observer (full name Guillaume-Joseph-Hyazinthe-Jean-Baptiste Le Gentil de la Galaziere) who studied deep sky objects, particularly nebulae. He presented results of his observations and discoveries about one month before Charles Messier discovered his first nebula. He traveled to see the transits of Venus in 1761 and 1769, a war preventing him from getting to the first and clouds preventing observation of the second.

Li Tian (c. Tenth Century). A Chinese monk who lived near the city of Liu Yang in Hunan Province and was credited with the invention of firecrackers.

Lowell, Percival Lawrence (1855–1916). An American businessman, mathematician, and astronomer most remembered for his description of "canals" on Mars. He founded the Lowell Observatory to search for intelligent life on Mars. His work at the observatory led to Clyde Tombaugh discovering Pluto after Lowell's death.

Magellan, Ferdinand (1480–1521). A Portuguese explorer who described the two large galaxies easily seen in the southern hemisphere sky. Hence they were named the Small and Large Magellanic Clouds.

Marco Polo (1254–1324). An Italian trader and explorer who is credited with bringing back Chinese gunpowder, and hence fireworks, from his adventures into China.

Maury, Antonia Caetana de Paiva Pereira (1866–1952). An American astronomer employed at Harvard College Observatory, where she observed stellar spectra and published a catalog of classifications.

Messier, Charles Joseph (1730–1817). A French astronomer most famous for his list of sky objects that could be mistaken for comets.

Morgan, William Wilson (1906–1994). An American astronomer whose principal work was the study of stellar and galaxy classification, notably in conjunction with Philip Keenan. (See also the entry for Keenan, above.)

Newton, Isaac (1642–1727). An English mathematician and physicist who is generally regarded as the founder of modern science. His brilliance spanned many disciplines of optics (the spectrum), mathematics (calculus), mechanics, and gravity, chemistry, and more.

Parsons, William, Third Earl of Rosse (1800–1867). An Irish astronomer who completed the building of a 72-in reflector at his home at Birr

Castle, Parsonstown, in Ireland, with which he detected the spiral structure of the Whirlpool Galaxy.

Pickering, Edward Charles (1846–1919). An American astronomer and physicist and director of the Harvard College Observatory. He produced a catalog of the brightnesses of stars assisted by women such as Annie Jump Cannon, Henrietta Swan Leavitt, and Antonia Maury. He published the first photographic map of the whole sky.

Planck, Max (1858–1947). A German physicist (full name Karl Ernst Ludwig Marx), renowned for (among many things) stating that electromagnetic energy could only be emitted in discrete packets, or quantized form. Hence, the quantum theory.

Plaskett, John Stanley (1865–1941). A Canadian astronomer and engineer who created a new spectrograph with which he measured the radial velocity of stars. His work contributed to knowledge of the rotation of the galaxy and location of its center. A binary consisting of two massive stars is named after him.

Pogson, Norman Robert (1829–1891). An English astronomer who computed the orbits of two comets and discovered five asteroids and six variable stars. He noted that stars of first magnitude were approximately 100 times as bright as sixth magnitude stars, which led to a redefinition of the scale of brightness.

Ptolemy V Epiphanes (204–181 b.c.). The author of the writings on the Rosetta Stone.

Pythagoras of Samos (c. 582–500 b.c.). A revered mathematician and philosopher who had great influence over mathematics, philosophy, and music, although none of his writings have survived to prove such achievements were his.

Rayleigh (1842–1919). Also known as John William Strutt, Lord or 3rd Baron Rayleigh. An English physicist who jointly with William Ramsay discovered argon. He also discovered the effect called scattering, that is, the dispersion of light by particles smaller than the wavelength of light. Scattering increases with decreasing wavelength.

Roche, Édouard Albert (1820–1883). A French scientist who formulated a theory of the gravitational effect on bodies circulating planets.

Russell, Henry Norris (1877–1957). See Hertzsprung above.

Schiaparelli, Giovanni Virginio (1835–1910). An Italian astronomer most noted for his observations of Mars and, in particular, misinterpretation by others of his description of straight markings as *canali* (see Lowell above). He also made the connection between meteor showers and the orbits of comets.

Secchi, Father Pietro Angelo (1818–1878). An Italian astronomer who pioneered astronomical spectroscopy. He created the first catalog of the spectra of over 4,000 stars and was the first to state finally that the Sun is a star.

Shakespeare, William (1564–1616). Probably the most famous playwright of all time, whose works often gave clues to the lives of people in his day, such as the type of household lighting.

Thackeray, Andrew David (1910–1978). A South African astronomer who first noticed dense, dark, opaque clouds of interstellar gas and dust set against rich star fields and glowing hydrogen gas, now known as Thackeray's globules.

Tombaugh, Clyde William (1906–1997). An American astronomer who became an assistant at the Lowell Observatory and discovered Pluto, a trans-Neptunian (dwarf) planet, that had been predicted by Percival Lowell. He never found any other planets but discovered many star clusters, clusters of galaxies, and around 800 asteroids.

Vaucouleurs, Gérard Henri de (1918–1995). A French astronomer who spent much of his time in America. He specialized in the study of galaxies and coauthored the *Third Reference Catalog of Bright Galaxies.*

Yerkes, Charles Tyson (1837–1905). An American businessman who funded the construction of the observatory of the University of Chicago in which is housed a huge refractor of 40 in aperture.

Young, Thomas (1773–1829). An English multitalented scientist also dedicated to the study of language. He is possibly most noted for his contribution to translation of the Rosetta Stone and the discovery of the diffraction of light.

Zwicky, Fritz (1898–1974). A Swiss astronomer born in Bulgaria who spent much of his time in the United States. He and W. Baade coined the term "supernova" and promoted use of the first Schmidt telescopes. He was the first to suggest that galaxy clusters could act as gravitational lenses.

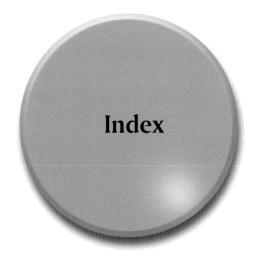

Index

Index

Printed in the United States of America